AF541086

History of INDIAN FISHERY

HISTORY OF INDIAN FISHERY

Dr. S.C. Agarwal
Director
Fisheries Department,
Haryana Government
Chandigarh

2018
Daya Publishing House®
A Division of
Astral International Pvt. Ltd.
New Delhi - 110 002

First Published, 2006
Reprinted, 2018

ISBN: 978-81-7035-426-0 (Hardbound)

Published by : **Daya Publishing House®**
A Division of
Astral International Pvt. Ltd.
– ISO 9001:2008 Certified Company –
4736/23, Ansari Road, Darya Ganj
New Delhi-110 002
Ph. 011-43549197, 23278134
E-mail: info@astralint.com
Website: www.astralint.com

"केशव धृत मीन शरीर जय जगदीश हरे"

"Matsya-Avtar temple at village Machchhyal, near Joginder Nagar, Himachal Pradesh on the bank of Machachyal tributaries of the river Beas. Fishing is completely prohibited in the vianty of 2 kilometers of this tributary near the temple. Mahaseer is the prominent fish in this tributary.

नमस्तस्मै गणेशाय यच्छुण्ड उत्प्लवायते
मत्स्यशास्त्राम्बुधेः पारं मेडल्पझस्य यियासतः

- डा. सतीश चन्द्र अग्रवाल

Prof. M.S. Johal
Professor of Zoology

FISH & FISHERIES LABORATORY
Department of Zoology
Panjab University
Chandigarh-160014 (India)
Ph. : 091-172-2534212 (O)
091-172-2709358 (R)
987-280-9358 (Mobile)
e-mail : johal_ms@yahoo.com
johalms@hotmail.com

"FOREWORD"

The History of Fishes in India is supposed to start with the publication of the book entitled 'An Account of the Fishes Found in the River Ganges And its Branches' by Francis Hamilton (Formerly Buchnan), M.D. from Edinburgh published by Archibald Constable Company, Edinburgh and Hurst Robinson and Company, 90 Cheapside, London England in 1822 for those who are engaged in the identification of the freshwater fishes. It will be interesting to know that various religious books in India dating back Vedic and Pre-Vedic periods pertain voluminous information on different types of fishes, various aspects of their fishery biology *e.g.*, food and feeding habits, reproductive biology, close season(s), muscle quality, fishing gears, angling etc. which is unknown to most of the Indian Fishery Workers probably due to the lack of documentation. Dr. S.C. Agarwal has done this arduous job by collecting the information available mostly in Sanskrit language from various sources, translating this information into English and included in this book entitled 'History of Fishes'.

My congratulations to Dr. S.C. Agarwal for bringing out this rare document, which will prove to be extremely useful to all the fishery workers in India and abroad. It serves the long-needed historical academic requirement on the subject of 'History of Indian Fishery'.

Hope the readers would find this book both interesting and rewarding.

Prof. M.S. Johal

"PREFACE"

Fish occupies an important place in human society at present and was also important in the past. In Indian literature fish is commonly known as Matysa, Meen, Machchali, Machchi, Machcha etc. in India. Fish is an aquatic cold blooded creature having fins and gills. It belongs to super class pisces of animal kingdom.

As per great epics 4 vedas, 18 purans and other literatures written in Braham, Sanskrit, Pali and Devnagrik scripts, and per Hindu mythology, there are more than 84 lakh living species on the earth. The description of aquatic species are also founds in the Vedic literature which was written most probably beyond to 10,000 B.C. As per Chark, Vedic Physician, there are more than 3 lakh aquatic animal species comprising about 30,000 species of fishes. The importance of fish in Vedic time can be judged from a Vedic book known as "Matsya Puran" describing the Lord Vishnu incranation in the shape of fish for the welfare of human race.

In addition, the fish provides food for man used as ornaments in the ancient times. The fish was also main source in paintings, handicrafts, sculptures etc. Fish as such has been used in various customs, traditions and ceremonies in different regions in the past and present.

I tried my best to consult the maximum literature to have the information on the importance of fish in the library of Archaeological Survey of India, Department of Archaeology, Haryana and Punjab University, Chandigarh.

The writings by Dr. S.L. Hora (1935-51), Dr. Hamid Khan Bhatti (1942), Dr. K.S. Mishra (1938), Dr. J.C. Gupta (1941) Dr. N.G. Majumdar (1934), Dr. D.D. Mookerjee (1935) and Dr. B.R. Singh (1998) have helped me in the preparing of this book.

I am thankful to Haryana Government for permission to publish the book.

I would like to use this opportunity to acknowledge the co-operation of my wife Sushi to spare me from house-hold activities so that I can write this book.

Dr. S.C. Agarwal
 # 3817, *Sector-22D,*
Chandigarh

"CONTENTS"

Foreword *ix*

Preface *xi*

Chapter 1: Introduction 1

Chapter 2: Archaeological Aspect of Fishery 6

Chapter 3: Fishery in Vedic Period 21

Chapter 4: Fishery in Medieval and British Colonial Period 57

Chapter 5: Fishery in Independent India 114

Chapter 6: Indian Fishery: A Future Vision 154

Bibliography 166

Species Index 171

Species Index 175

Chapter 1

"INTRODUCTION"

Every drop of water contains life. Many drops form the aquatic system. More than 2/3 area of the earth is covered with water. It is believed that life began in water first on land. Fish lived in aquatic environment for the more than 350 millions years since the Devonian times. The fish are the most numerous of all vertebrates. There are more than 22000 species or so outnumbered the grand total of all other vertebrates classes taken together. Broadly, fishery means business of catching fish or industry of fishing. The fishery concerns itself with the habits, life histories and inter-relationships of fish population. It also concerns to the effects of commercial fishing on its population, and inter-relationships of various species of fish to each other. In the ancient water-oriented civilizations, fishing was probably the earliest form of hunting. Man was surely fish hunter before cultivator. Fishing is probably the oldest industry in the world.

The history of biology by Nordenskiold (1929) records that the civilized people of eastern Asia, Hindus and Chinese, have contributed little for understanding of fish and fisheries science. Such an impression among western scholars seems to have resulted from paucity of information on the achievements in the field of Biological Sciences in India.

Evidences of fishery in the Indus Valley and the adjoining areas of Sind, Punjab and Baluchinstan have been brought out by a careful study of the actual remains of fish and fish hooks found in Mahanjodaro, Harappa, Amri, Nal, Nundara and Rupar cultures. Seals, amulets, figurines, paintings on the pottery and other wares have been engraved with fish designs. The designs consist of fishes belonging to the fresh water genera, *viz. Labeo, Rita* and *Wallago* and to the marine cat fish of the genus, *Arius*. Fish are connected with Hindu legends. The movement of the Sun and its relation to the change of season was of paramount importance to the agricultural people. The configurations of the fixed stars were grouped to constellations which were given names of animals and mythological figures. An imaginary band the "Zodiac", also called the circles of animals was established to mark the movement in the sky. It was divided into 12 equal parts each called a "house". The twelfth house was called "Pisces" which was illustrated by two fishes tied together by their tails. In Hindu Astrology, 'Pisces' or 'Meen' Rashi governs the stars of a certain percentage of people under its influence. It enjoys an elevated position in Hindu mythology and stands as a symbol of good omen or "Subh".

मीनो पुच्छास्य सलग्नौ मीनराशि दिवाबली,
जली सत्वगुणाद्‌यश्य स्वस्थो जलचरो द्विनः।
अपदो मध्यदेही च सौम्यस्थे लयुभयोदयी,
स्‌रा चार्याधिपश्चास्य राशीनां गदिमा गुणाः
गिश्दाभागत्मकानां च स्थूल सूक्ष्मफलाय च।।

Fish is also a symbol of Vishnu or Budha. Among Hindus, it is an incarnation of God Vishnu:

"केशव धृत मीन शरीर जय जगदीश हरे"

Fish occupies an important place in the Indian mythology, history and tradition. For instance, one of the incarnations of God was in the form of fish, "Matasyavatara". The avatars referred to in Hindu scriptures are associated with common animals. These are arranged in chronological order of evolutionary significance from the fish and the turtle, through the boar and the lion-man to the pigmy-man in the first five of the avatars.

Animal world, including fishes, was not a separate entity to the Vedic people, but a part and parcel of the great cosmic system embracing the whole world. Biological phenomena are therefore, diffused throughout the Vedic and Post-Vedic literatures. Fish has been mentioned in Rig-Vedas and later more frequently in Adharva-Vedas, Yajur-Vedas, the Samhitas, the Upanishads etc. "Jala" or fishing net has also been mentioned in the Atharve-Vedas, Sutras and the hook in the Yajur-Vedas. As evident in the Susruta Samhita, the knowledge of inland waters and fresh water fishes and fishermen occur in the epoch-making books like Mahabharat and Ramayan. Kalidas also described the fishermen and their profession in one of his master piece work, Abhigyan- Shakuntalam.

"जम्बू फलनि पक्वानि पतन्ति विमले जले।

तानि मत्स्य न खादन्ति, जाल गोलक शंकया।।"

The value of identification of animals was not unknown to the ancient Indians. The nomenclature adopted by them was generally derived from the characteristics of the animals or their habitats. The ancient Hindus classified the animals into groups which even hold good to-day. For instance, the animals were divided into attinika (having a back bone) and anattinika (without a back bone) which correspond to the modern classification to vertebrates and invertebrates. The number of feet was used to divide animals such as Dipada (Bipeds), Chatupada (quadrupeds), Chapadda (hexapeds) and Satapada (Centipeds). The animals were also classified according to their habits and habitats. For instance, the ancient Hindus recognised Jalacocera (acquatic), lacocera (terrestrial), Ghara (domestic) and Arannaka (wild) animals. Thus the ancient Hindus had a fair knowledge of classification, bionomics and habitats of the animals. It must be emphasized that Charak, Parasatapada, Susruta, Sankara and Umasvati classified animals on rational basis. These classifications, are in no way inferior of those of Aristotle, Palini or Gesner.

The economic managements of fishes were laid down by Ashoka on pillar edict V in 246 B.C. and in Kautilya's Arthasastra about 300 B.C. The record of fish farming in Bengal can be traced back to the Pal Dynasty. It is evident from a copper inscriptions of Devapal (810/850 A.D.) in which a village moshika was made a gift with all conceivable right of cultures.

"स्वमीमा तृणयुतिगोचर पर्यन्त सतंत्र सोद्देशः
साम्र मधुकरः सजलस्थल समत्स्यः सतृण"

(Mention of three distinct crops are made here-mango, mahua and fish).

It is well known that fish decomposes rapidly, particularly in warm climates. This is particularly due to activities of the bacteria and moulds that it harbours, and of enzymatic auto-digestion and atmospheric oxidation. From ancient times, preservation by drying, salting or smoking have been developed and handed down from generation to generation. The methods of fish curing have permitted storage of day to day and seasonal surpluses until times of dearth. Its impact on cultural development of the human race was considerable. The food habits and food patterns of man have changed with the time. Most important was the abundance of food and the possibility of storing it, which permit the development of larger communities and the occupation of sites that provided safety and shelter to them for a considerably longer period than had been possible before.

Bones of marine fish found in Mohenjodaro communities which lived on the river Indus during the Bronze age period (5000 B.C.), clearly indicate that trade including sea fish products was established between the coast and inland settlements. References of fish trade and fishermen communities are also found in the songs of Sangam age (First to Fourth century A.D.). The fish has appeared in the folk of fishing communities of that time. The sailors have continuously added new fish tales to the old ones.

Fish is considered to be an auspicious object by the Bengalees and Bhiarees in most of their festive occasions and social practices like worships, marriage ceremonies, Anaprasna, wearing of sacred thread, rituals, enjoyments, presentations and others. It forms an indispensable part in the life and living of the Bengalees since time immemorial.

Fish occupies a place of importance in our fine arts. Fish designs are very frequently used by house-wives in their 'Alpana', 'Brata' designs of patched clothes, models, mould of sandesh and khir. The toys, artificial leaves and flowers are also made from fish scales. Ornaments for women are engraved with fish disings by goldsmiths.

The place of fish in our culture would not be complete without having a least glance at sport fishing known as angling. It is the art or practice of the sport of catching fish by means of a baited hook or 'angle'. The sport of catching fish is an age old activity. It can be said to date from the time when man was making out a precarious existence by the slaughter of any living thing which he could reach with the rude weapons at his command. In the earliest literatures references to angling are not numerous. In the Old Testament, there are passages which show that fish tackling with hook was one ot the commonest industries in the East. It seems probable that attack on fish was at first much the same as attack on animals. Moreover, angling was a great source of recreation in streams, rivers, lakes and coastal waters in India. Matsyavinoda written in 1127 A.D. is the first known treatise on the art of angling in the world.

In Medival period and British colonial period the stress was raid on the conservation of fish in natural waters. In modern time, development of fishery has been changed to culture side rather to conservation to cater the high demand of fish and water products to support expending population.

Chapter 2

"ARCHAEOLOGICAL ASPECT OF FISHERY"

It is generally not recognized how fisheries have played in the civilization of man and in the destinies of the nations'. A full history of fisheries and even of the fisheries of a particular country has yet to be written. It is quite likely that when the first man ventured to sea in a dugout, his purpose was to pursue fish. Not only the modern deep-sea trawlers, liners and drafters but the merchants and fighting navies are the liberal descendants of the prehistoric fisherman in his dugout.

That anthropologists use the concept of culture to cover all those skills and ways of life that transmitted by inter-personal communication and tradition rather than by genetical means. Our knowledge of history of man's earliest culture is almost entirely restricted to the stone tools that have been preserved.

The sequence of industries was first worked out by archaeologists in Europe. The most widely used names for the different stages of man's culture are derived from the places at which the industries were first discovered. The sequence of industries is called Paleolithic to cover the whole period during which instruments were made by chipping and the people were hunters and food-gathers. The stages of culture were not reached every where at the same time. This makes the difficulties in nomenclature and dating.

The concept of three Ages *viz.* Stone, Bronze and Iron was first given by the Danish Archaeologist, Vedel-Simonsen in 1813. The stone age was later split up into Paleolithic, Mesolithic and Neolithic cultures. The time periods for these cultures have been determined by Clark (1969) to be around 5,00,000–35,000 BC for the Paleolithic, 11,000 BC for the Mesolithic and 9000 BC for Neolithic culture.

Paleolithic Culture

Stone tools of the Paleolithic period have been found distributed all over India. The man in the Paleolithic age was supposed to be vegetarian. He became carnivorous at a later stage and took to hunting and fishing when he began to make tools. In India, Paleolithic tools have been discovered in the neighborhood of rivers, small streams and lakes. The stone-age man used to hunt for the wild animals that lived in jungles along the rivers and lakes. When the animals came for drinking water, the Paleolithic hunters had an opportunity to kill them. Apart from the wild animals, fish found in lakes and rivers must have attracted their attention and formed the source of food for them.

Mesolithic Period

Mesolithic period (5000–3000BC): Mesolithic man usually used the microliths to point and barb their arrows, knives and also scrapers for hunting and food gathering. Microliths have been found in large quantities from the Indian subcontinent, Sankalia (1962) pointed out that these microliths must have been made by the ancient man for the purpose of hunting and fishing as they lived in temporary camps near the river banks or sea–coasts. He stated further that the people of Gujarat (Langhaj) who lived in the sandy undulating plains of northern and central Gujarat were proficient in manufacturing microliths.The Langhnaj people occupied elevated areas of sand-dunes formed near lakes and lived by hunting and fishing. This becomes evident from large quantities of bones of cattle, nilgai, deer, rhinoceros, squirrels, rats, tortoise and fish found in their habitations. Arrow heads made of stone indicate that they used bows and arrows for hunting and fishing.

The best evidence of life and activities of Mesolithic man lies in the paintings in cave-shelters located in India. The most exciting stone age paintings are from Madhya Pradesh discovered by Wakankar (1975). In the valley of the Sone river and the Mahadeo hills and Singhanpur, rock paintings discovered have been compared with Paleolithic paintings in Europe.

Neolithic Period (3500 B.C.–1700 B.C.)

Neolithic period in India is difficult to determine. The transition from hunting to farming way of life in the Neolithic culture took place gradually. Flying microliths were used to cut crops and vegetables. It began in South-West Asia around 9000 B.C. and spread gradually through Asia to China. Neolithic man could smelt copper and lead and used them for beads and ornaments but not for tools.

In Bronze age (5,000 B.C.), metallurgy grew for making of copper axes and silver vessels to the addition of tin to copper to produce the full Bronze Age Technology.

The discovery of iron some 2000 years later (Iron Age) accelerated the process of change that metal produced. With these materials, the inventive capacity of the human brain soon began to release the stream of new aids to life that has increased until today. The discovery of new tools and its use to fight for increasing necessities for success.

Pre Harappan Peasant Communities of Baluchistan and Fish Paintings

From the pre-harappan peasant communities of Baluchistan some pottery remains with paintings have been found. On the basis of techniques employed in pottery paintings, a broad classification of the pre-harappan Baluchi cultures has been proposed as follows:

- (A) Buff-Ware culture:
  - (1) The Quetta culture:
  - (2) Amri-Nal culture (from two sites, the first in Sind, and the second at the Nal–Valley in Baluchistan).
  - (3) The Kulli culture
- (B) Red-ware culture:
  - (4) The Zhob culture

Some interesting pottery remains with fish paintings have been excavated from Nal (Jhalwan division of Kotat State in Baluchistan).

Location of Nal

Nal valley in Baluchistan is about 30 miles long and seven miles broad. It lies in Lat 17°40′N. Long 60°48′E and is nearly 3834 feet above the sea level. The pottery discovered from here is invariably wheel made. The animals represented are fishes, birds, scorpions, bulls etc. The various types of pottery found are known to be funerary in character as evident from their association with burials. The Nal type of pottery has also been found from other places like Nundara in Baluchistan and Damb Bhuti in Sind.

Data available on fish paintings from Nal have been brought together and discussed by Hora (1955). From the fish decorations on Nal pottery, there can be little doubt about their general identification except for a pair of fishes painted on Vase No. 142 of Hargreaves (1929) list. However, their identification at the species level cannot be said to be certain. Different genera and species of fish representing the paintings were identified by Hora (1955) as follows:

Vase No. 115 and 140 of Hargreave's List

The fish painted on this vase strongly resembles with the genus *Garra* (Ham.). It is well known that the different species of the genus *Garra* contain much oil and are greatly relished by hilly people. It seems possible that the people of the Nal culture considered Garra as a food fish of great value and hence painted it on funerary vases.

Vase No. 142 of Hargreave's List

This painting seems to represent some species of the genus Cyprinion (Heckel), many of which have been recorded from Baluchistan (Berg, 1933).

Vase No. 184 of Hargreave's List

The illustration represents a hill stream fish, probably *Labeo* spp. The representation of scales differs considerably from other Nal fish drawings.

Vase No. 195 of Hargreave's List

These paintings represent more accurate representations of fishes with scales and fins. The two fishes seem to represent *Crossochilus latus* (Hamilton) and *Labeo dyocheilus* (Mc Clelland). Both the species live in the hill-streams all along the Himalayas and constitute the important food fishes of the region.

Vase No. 193 & 196 of Hargreave's List

Five fishes have been painted around this vase. They can be readily identified as the golden Himalayan Mahser or *Barbus, Tor putitora* (Hamilton). Hora (1955) was of the opinion that certain special features of this Mahseer as shown on the Nal paintings particularly in regard to the variations in the form of the snout were not due to any fancy of the painter, but were correct representations of what happens in nature. It is remarkable that such differences were not overlooked by the painters of the Nal pottery.

Mahseer is considered superior to *Rohita* and *Pathina*. In the Hindu Dharma-Sastra, it is customary to believe that if Brahmisn are fed on Mahseer's, the men remain stratified for ever (Hora, 1953). Even hermits are advised to build their huts at places where Mahseer is available. The value of Mahseer as a food fish seems to have been well realized by the people of Nal culture.

Vase No. 256 of Hargreave's List

Four pairs of different genera and species of fish are painted on this vase. These include the following:

1. *Tor putitora* (Hamilton)–large scaled barbells, head longer than the depth of the body, large scales diagrammatically represented by cross lines.
2. *Nemachilus* sp.–painted as banded loach, colouration seems to represent *N. baluchiorum*.
3. *Botia* sp.- painting resembles *Botia dario* (Hamilton) except for the colour bands which are shown to run in reverse direction.
4. *Glyptothorax* sp.–painting shows a pair of cat-fishes in which the body is not covered with scales. The depressed head and body alongwith the flattened ventral surface are similar to the species comprising the genus *Glyptothorax* (Blyth).

From the decorations of the hill stream fishes on Nal pottery, it is evident that fast-running clear streams existed in its vicinity. Except for the genus *Cyprinion*, fishes of the other seven genera out of eight painted (*Garra, Labeo, Crassochilus, Tor, Nemachilus, Botia, Glyptothorax*) are well represented in the perennial streams of Himalays. Some of the genera like *Garra, Labeo, Tor* and *Nemachilus* have extended their range to Africa as well. *Glyptothorax* is a fish of torrential stream. Its occurrence at the time of Nal painting is indicative of a heavier rainfall and a series of perennial streams existing at that time.

The doubtful recording of the genus *Cyprinion* from Nal painting is very significant. This genus represents a western element in the fauna of pre-partition India. Species of *Cyprinion* have been recorded from Syria, Mesopotamia, Persia, Baluchistan, Sind and Punjab. Six species of the genus are now known from Sind and Baluchistan. The genus *Scaphiodon* was established by Zugmayer (1914) and sytematised as *Cyprinion* by Berg (1933). The interesting point that not a single species of the genus *Cyprinion* was properly painted on the Nal pottery points to the probability that this genus was not well established in Baluchistan during the time of the Nal culture. The Zoogeographical distribution of the six species is as follows:

1. *Cyprinion watsoni* (Day)–Sind
2. *C. watsoni belensis* (Zugmayer)–East Persia and Baluchistan.
3. *C. irregularis* (Day)–Persia, Baluchistan, Sind, Salt Range and S. Waziristan.
4. *C. microphthalmum microphthalmum* (Day)–S. Persia, Seistan and Baluchistan.
5. *C. microphthalmum nikolskii* (Berg)–S. Persia and Baluchistan.
6. *C. milesi* (Day)–Southern Persia and Baluchistan.

Zoogeographical distribution clearly indicates that the species are found in Baluchistan and also found in Persia. Thus, it seems logical that in the prehistoric fish fauna of Nal, the westward movement of the Indian fauna stopped and eastward movement of the Middle East species commenced.

The decoration of fish on Nal pottery is testimony of the fact that the hill stream fishes were relished as an item of diet. The painting of the genus *Glyptothorax* (Blyth) is significant. Though this fish is found in Persia on the one hand and throughout the South-East Asia on the other. It is not recorded from Baluchistan hills. Its painting on Nal pottery possibly helps us to bridge the gap between India and Persia to some extent. Hora (1955) rightly concluded that Baluchistan, now an arid area might have had more rainfall and a number of perennial streams during the time of Nal culture as evident from at least three of the fish motifs on the pottery represent hill stream genera.

Methods of Fishing at Nal

Since no fishing implements have been excavated from Nal, it can be inferred that the usual methods of fishing by nets and angling were not applicable to torrential streams. Then the methods of fishing must have been damming small channels and allowing them to run dry or by poisoning the pools with certain plant juices which stupefy the fish and cause them to float on the surface. Such methods of fishing though considered illegal, are still practiced at many places in the hills. The long survival of traditional methods of fishing in the hill areas suggests that the techniques now in vogue are probably derived from those known during the Nal culture.

Fisheries in Harappan–Chalcolithic Culture

The central point of Harappan–Chalcolithic culture is the Western Punjab. This civilization slowly diffused to Bahawalpur (Eastern Punjab), Haryana and Jammu in about a century. The antecedents of the Harappa culture are the village sites of Baluchistan hills, the Nal culture and the Kulli culture from the Makran coast to the west of the Indus delta.

The Indus valley civilization or the Harappan culture is older than the Chalcolithic culture. The Indus valley civilization is also called Harappan civilization because it was discovered at the modern site of Harappa situated in the province of West Punjab in Pakistan in 1921. The Harappan culture covered parts of Punjab, Sindh, Baluchistan, Gujrat, Rajasthan and Western Uttar Pradesh. The Harappans produced their own characteristics fish hooks, pottery and seals.

Fish Remains of Harappa Culture

Fish remains obtained from the excavation at Mohenjodaro, Harappa, Ropar, Rangpur and Lothal belonging to Indus Valley culture indicate the following types:

Mohenjodaro

1. *Rita rita*
2. *Wallago* sp.
3. *Arius* sp.

Harappa

1. *Rita rita*
2. Remains of Carp.

Pot-shreds from Harappa

The most remarkable post-shred from Harappa shows fisherman carrying two nets suspended from a pole across his shoulders with a fish and probably a turtle near his feet. These rest on a cross hatched band, presumably a river by which he is walking. Fragments of another pot show a fish and star in the upper part of the panel round its circumstance and secondly a man with one hand raised and the other touching his head, a child with upraised arms with two fishes and a cock in the field. Vats (1940) has pointed out that fish toys were also found from Harappa.

Fish Hooks from the Indus Valley

The art of angling seems to have reached a high level of perfection in the Indus Valley civilization. Apart from the Indus Valley, none of the others excavated or explored site has yet brought to light any specimen of fish hook in India. At the same time, the aboriginal tribes of India appear to be ignorant of the method of fishing by hooks. The Maria Gonds (Grison, 1938) use horns as hooks to catch fish. The Sakai of Malay-Pennisula (Skeat and Blagden, 1906) use curved thorns of rattan (Calammus) as fish hooks. Radcliffe (1921) has brought to light a natural fish hook from new Guinea made of the thick upper joint of the hind leg of an insect, *Eurycantho latro*. Fish hooks made of turtle shells or wood are widespread among the aboriginal people of the pacific Island, North America and East Africa. Lagercrantz (1943) has shown that there are two centers of indigenous fish hooks in Africa; one in E. Africa and the other in Congo. No such indigenous fish-hooks are met with among the aborigines of India. Fishing by gorge-hooks is however, practiced in the case of Koi (*Anabas*) fishes in paddy fields of East Bengal. It is probably one of the local indigenous methods as it is not prevalent in West Bengal.

The fish hooks of the Indus Valley can be classified into two main types: (1) Barbed and (2) Unbarbed. Specimens of 16 fish-hooks from Mohenjodaro have been mentioned by Marshal (1931) and Makay (1938) and one from Harappa. Excavation at Chanhudaro has yielded 7 specimen including Majumdar's collection (1934). All the 16 specimens of Mohenjodaro are barbed. Of the Chanhudaro finds, 3 are barbed and 4 are unbarbed. The solitary specimen from Harappa is unbarbed. The paucity of fish-hooks at Harappa indicates that angling was not popular at that place.

The variation in the relative sizes of the Indus Valley fish hooks evidently shows that hooks of different sizes were employed for catching different kinds of fishes. Mohenjodar specimens vary between the lengths of 2-9" and 1-04". Chanhudarao hooks vary between the lengths of 1-87" and 0-43". The difference in the relative sizes of the hooks indicate that fishing was probably confined to tanks and small streams at Chanhudaro whereas at Mohenjodaro all fishing was done in the Indus. Apart from fishes from rivers and tanks, the Indus Valley people also used fishes belonging to the estuarine forms of the genus *Arius.*

The evolution of barbs at Chanhudaro deserves special attention. It appears that the barbing process underwent some experimentation at this place. It is evident that the hook No.3038 has a barb at the center of the hook, whereas hooks Nos. 2943 and 3641 have their barbs made by a V-shaped fork at the end. The final and perfect form is seen in hook No.3487. It seems probable that the fish hooks of Indus Valley reached their final and perfect form at the 4th stage.

1. Barbless
2. Barbed near the center
3. Barbed towards the end
4. Barbed inwards at the lower half

It is of interest to note that Chanhudaro has yielded the only form of a hook with a central barb.

Fish hooks with barbs opened outwards at the end have been reported from Africa as well (Lagercrants, 1934). The similarity between the fish hooks from Africa and Chanhudaro is striking. There is found two fish hooks with end barbs. Their similarity with the Chanhundaro finds is only in having the barb at the end. The actual barbs are different.

So far as the other end of the hook is concerned, the Indus Valley civilization has yielded more than one variety. Mackay (1938) holds that it generally consists of a straight shank slightly thinned out and turned over to form an eye at the top. Three distinct types of hooks can be clearly distinguished. These are:

1. Hooks consisting of a straight shank without forming any eye.
2. Hooks consisting of straight shanks but turned outwards; the bent loop does not tough the shank.
3. Hooks consisting of a straight shank but curved out to form a small eye. The eye is either connected with the shank or only slightly separated from it. There is one specimen in which the eye is turned sideways.

Lagercrantz (1934) has shown a long gap between the barbed and barbless hooks found in Egypt. The presence of both the barbed and barbless hooks in the Indus Valley together with the variation in their relative sizes seems to be an independent character of the Indus Valley culture.

Urs a historian has produced two examples with end barbs. The Urs specimens (Woolley, 1934) differ from the Indus Valley specimens in the end of the hooks. The peculiar hook with a central barb is undoubtedly similar to that of No.2943. It's final form is evidently No.3487. Another find from Urs (Hall and Woolley, 1927) has some resemblance to certain specimens from Mohenjodaro, so far s the barb is concerned. However, its workmanship appears crude.

It appears form the above facts that the best type of metallic fish-hooks was probably developed at the Indus Valley. In fact, it attained the best perfection of all the fish-hooks found in Egypt and Mesopotamia. The agreement between the hooks discovered earlier and the modern fish hooks may be the continuity of a culture-trait as found in pottery designs by Mackay (1930).

In eastern India, fish hooks were found in Bihar and West Bengal, where rice was also found. This suggests that the people belonging to the stone/copper age in eastern region lived on fish and rice, which is still a popular diet in this part of the country (Sharma, 1986).

Attempt has been made by Sarkar (1954) to identify the other artifacts particularly boats which might have been associated with fishing. It cannot be denied that although the majority of the Indus Valley sites are situated on the banks of navigable rivers, yet artifacts connected with boats and navigation from the archaeological finds have not been discovered.

Circular stones of various sizes (½″ to 4″ in diameter) with central hole have been found at Harappa and Mohenjodaro. Such stones are conspicuously absent from the sites at Chanhudaro. The large ringstone with heavy weights might have been used as anchors for boats.

Fishing nets have been identified from paintings on a pot-shred found at Harappa. It showed a fisherman carrying two nets hanging from a pole across his shoulder. Some fishes and turtles are lying close to his nets. The fisherman stands on a cross-hatched band possibly a river as identified by Vats (1940).

Net-sinkers discovered from various sites may be classified into two types, namely (i) hollowed or annular and (ii) grooved. While type I has been found from Mohenjodaro, type II from Mohenjodaro and Chanhudaro both. Reports from Harappa do not contain any reference to net-sinkers. Type I is similar to pottery-rings found along with the remains of fishing net from Khafaje, a place nearly 10 miles east of Baghdad. Type II resembles A1 'Ubaid net-sinkers (Hall & Wooley, 1931) which were made of rectangular pieces of pebbles with a groove round the center.

A large number of slab-like objects have been unearthed at the Indus Valley sites. Their probable use as net-weights cannot be ruled out. Chanhudaro has yielded a large number of such objects. They all appear to be net sinkers. Small ring-stones may also serve the purpose of net weights. Fishermen of the sea coast of Puri (Orissa) frequently use stone blocks as net-sinkers (Bose, 1954). Stone weights in nets is known to be used by the Egyptians (Radcliffe, 1921). Stone–weights to anchor the narrow end of a net is practiced in Denmark and Sweden also (Clark, 1948). The Indian fisherman frequently use slabs of stone of bricks with dra-nets.

In the cast, 'beads' serve the purpose of net-sinkers. The cylindrical, spherical, elliptical or even barrel-shaped beads classified by Beck (Vats, 1940) might have been used as net sinkers. The net sinkers in India are usually made of cast iron. Their sizes vary according to the type of the net to be used. Hornell (1924) has classified two types of terracotta net sinkers, namely (i) drum-shaped and (ii) annular. An annular net sinker about ¼″ in diameter collected from Rangpur, North Bengal has been kept at the Ethnographic

Gallery of the India Museum. It conforms to the hollow type of net sinkers discovered from Mohenjodaro. A large number of pottery rings, about an inch in diameter; have been abundantly found in the Indus valley. Their employment as net-sinkers has been rightly considered by Mackay (1938).

Boats in Ancient India

The origin of the word 'boat' is probably to be looked for in a piece of wood. If this be so, the term in its inception referred to the material of which the primitive vessel was constructed. We may assume that man in his effort to achieve the feat of conveying himself and his belongings by water succeeded in designing a boat made up of wood.

We get direct evidence of boats being used for navigation in India. These evidences can be collected from the Indian art, sculptures, paintings and coins, besides these, Sanskrit and Pali literature furnish few references which have a direct bearing on the ships and ship building in India. The account of vessel used by the Indian generally refers to a large craft which naturally falls into the head of ship. There are several representations of boats and ships in old Indian arts. The earliest of these is the one found on a rectangular seal unearthed by the archaeological excavation as Mohen-jodaro in the Indus Valley dated to at least 3000 B.C. The absence of a mast suggests that this boat was used only for the river work. According to Mackey Mackay (1938) a boat with mast, scratched on a pot shred was also found in which the steersman is shown at the opposite end. This indicates that the boat is sailing in the other directions from the boat on the seal. The boat resembles the first being high at both ends and having also a steersman. Instead of the deck cabin in the center, as in the other boat, it shows a mast, very probably, the second type of boat or vessel for both riverine and sea trade.

Other early representations of boats and ships in Indian arts are to be found among the Sanchi sculptures of the 2nd century (B.C.). One of the sculptures on the Eastern gateway of No.1 stupa at Sanchi represent a canoe made up of rough planks rudely sewn together by hemp or string. "It represents river or a sheat of freshwater with canoe crossing it, and carrying three men in the ascetic priestly costume, two propelling and steering the boat, facing towards four ascetics who are standing in reverential attitude at the water edge below".

There is another sculpture to be fond on the western gateway of No.1 stupa at Sanchi which represents a piece of a water, with a cargo boat floating on it whose prow is formed by a winged gryphone and stern by a fish's tail.

Among the sculpture of the temple of Lord Jagannatha at Puri a fine well preserved representation of royal cargo boat shown in relief on stone deserves special mention. The representation appears on that portion of the great temple of Lord Jagannatha which is said to belong to the 12th century A.D. The sculpture shown in splendid relief is stately barge propelled by lusty oarsmen with all their might.

In Bhubaneshwara there is an old temple on the west side of the Tank of Vindusarovara also called Vaital Deul after the peculiar form of its roof resembling a ship or boat capsized. The word vaitara denotes a ship.

There are few representations of old Indian boats and ships among the famous painting of the Buddhist cave temples at Ajanta. The execution of these works is supposed to have extended from the 2nd century B.C. to 8th century A.D. covering a period of more than a thousand years.

References of Boats in the Sanskrit and Pali Literature

The book called "Yukti Kapatau" a compilation by Bhola narapati, is some sort of a treatise which elaborates the art of the boat and ship building in ancient India and sets forth many interesting details about the various sizes and kind of ships, the materials out of which they were build and the like. It also sums up in a condensed form all the available information and knowledge about the then ship building industry in India. The boat and ship builder in ancient India had a good knowledge of the wood materials as well as varieties and properties of wood which they used in making boats and ships.

The Pali texts do not provide the actual measurements of the different dimensions of ships as the Sanskrit texts furnish. According to Rajavalliya, the ship in which prince Vijaya and his followers were sent away by kind Simhava Bengal was so large as to accommodate full seven hundred passengers, all Vijaya's followers.

Boats in the Vedic Period

Nav or Nava is the regular word used in the Rig-veda to denote a boat or ship. PLAVA (Float) denotes a boat in the Rig-veda and in the Jaittiriya Samhita. Vau is also referred to in the vedic literatures Jaittiriya Samhita, Vajasneji Samhita, Aiteryea Brahmana, Satpatha Brahmana) for a boat or ship.

Thus, in the Vedic ages, shipping was in infancy and fairly large boats served the purpose of navigation in the rivers. It gets support from some of the terms used in the Vedas. These are:

Dyumna–(Raft)

It denotes raft in one passage of the Rig-Veda.

Aritra–(Oar)

It denotes the Oar by which boats are propelled.

Aritr

The rower of a boat is called aritr.

Navaja

Refers to a boat man in the Sathpatha Brahmna.

In the Rig-Veda it has been referred that the Asvins rescued Bhujya in the ocean with a ship of a hundred oars (Sataritra). Moreover, it is also mentioned in the Rig-Vedas that men used to go to ocean (Samudra) for gain (Sanisyavah).

According to Vriksh-Ayurveda, four different kinds of wood are to be distinguished–which go in for making boats and ships.

In chapter VIII of Manusmriti, there is a passage that lays the rule of hiring a boat in case of river journey and sea voyage.

दीर्घध्वनि यथादेशं यथाकालं तरी भवेत् ।

नदीतीरेणु तद्विघात् समुद्रे नास्ति लक्षणम् ॥

Boats and Ships in the Epics

In Mahabharat the righteous Pandava brothers on the friendly advice of kind hearted Vidura escaped from the destruction planned for them in a ship. Vidura kept a ship ready and constructed for this purpose. The ship was provided with all necessary machinery and weapons of war.

तत्: प्रवासितो विद्वान् विदुरेण नरस्तदा ।

पर्थानां दर्शयामास मनोमारूतगासिनीम् ।।

सर्ववातसहां नावं यन्त्रयुक्तां पताकिनीम् ।

शिवे भागीरथीतीरे नरैर्विश्रम्भिभिः कृताम् ।। आदिपर्व्व ।

It has been mentioned in Ramayan that in five hundred ships, hundreds of Kaivarta young men were asked to be in wait and obstruct the enemy's passage.

नवां शतानां पच्चानां कैवर्त्तानां शतं शतं ।

स्न्जद्वानां तथा यूनान्ति ठन्त्वित्यभ्यवोदयत् ।।

(Ayodhya Kandam, 34.78).

Kalidas in Raghuvansa (Canto 4 sloka 36) describes that it was in ships that the people of Bengal once made a stand against the invincible prowess of Raghu who retired after planting the pillars of his victory on the isles of the holy river.

बंगान् उत्खाय तरसा नेता नासाधनोद्यतान्।

निचखान जयस्तम्भं गंगास्त्रोतोऽन्तरेषुच ।।

Alexender's passage of the Indus was effected by means of boats supplied by native craftsmen. A flotilla of boats was also used in bridging the difficult river of Hydapes.

Chapter 3

"FISHERY IN VEDIC PERIOD"

Early Vedic Period (1500 B.C.–1000 B.C.)

It seems logical that about 1100 B.C., a large wave of Aryans entered India. They moved slowly from North-west to east and the use of iron spread are found from Baluchistan, Gandhara to eastern Punjab, Western Utter Pradesh and Rajasthan. The Rig-Vedas-Samhitas is considered to be the earliest work of the Aryan race. The works throw light on the mode of hunting of fishes and feeding habits of people. The Vedic Aryans appears to be meat eaters rather than fish eaters. However, 'Matsya' or fish has been mentioned in the Rig Veda. Muksija is also found in one passage of the Rig Veda which clearly means a "net" for catching fish or animals.

The Later Vedic Period: (1000 B.C.–600 B.C.)

The later vedic works such as, Atharva Veda, Yajur Veda, Samhitas, the Brahamanas, the Aranyakas and Upanishads took place during the period 1000 B.C. to 600 B.C. The collection of the Vedic hymns or Mantras were known as the Samhitas. The hymns of the Rig Veda were set to tune. This modified collection was later known as the Sama-Veda Samhita. In Atharva Veda and Sta-patha Brahmana, reference to fisheries has been made by narrating a legend in which a fish saved Manu from deluge. The Vedas, the Samhitas, Brahmanas, the Aranyakas and Upanishads contain several references of fishres.

Fish names appear in these texts. However, these names may refer to individual species or to a genus as a whole. For example, the terms like Purikaya, Jasu Matsya and Sakula may refer to some genera or species of fishes.

The Post Vedic Period

In this period, the fishes were caught by nets, hooks, hands or by damming streams and baiting with poison as evident from the various names mentioned in Yajur-veda.

Sayana (Taittiriya Brahmana) says that Dhaivara is one who takes fish by netting a tank; Desa and Sauskala do so by means of fish hooks (Badisa); Kaivarta and Mainala by means of net (Jala); Margara by means of hands; Anda by putting pegs in dams and Parnaka by putting a poisoned leaf in the water.

Fish and Fisheries in Sutra and Smriti Literature (600 B.C.–200 A.D.)

After the Vedic period, separate treatises were written which framed rules concerning social laws, rituals and domestic ceremonies. These were respectively called Dharma-Sutras, Saruta-Sutras and Grhya-Sutras. The Smritis deal with the same subjects but in a more elaborate way. It gives more prominence to social laws. Hence, it is also known as the Dharma-Sastras.

We find only causal references to fish in the Sutras and the Smritis. Nevertheless, they reveal to a considerable extent the part played by fish in man's affairs in those early days in India. It also shows that for the Hindus, most of whom are now vegetarian, fish formed an important item of food at that time.

References to specific types of fish have been made in different contexts particularly either in connection with the 'Sraddhas' (*i.e.*, ceremonies wherein food is offered to the manes but is actually consumed by Brahmanas priests) or with certain forbidden foods. When not forbidden, certain species are shown to have edge over the others, either for higher merit or for their greater food values. Again, where mean formed a part of offered food, fish is often declared superior to flesh of other animals.

The most celebrated Smriti is that of Manu. It is also called Manu Smriti, Manava Dharmasastra or Manviya Dharmasastra. Parts of this work are assigned to 400 B.C. The work is a fairly extensive one, comprising nearly three thousand verses. In its chapter on the trans migration of soul, fish is bracketed with those creatures which are supposed to be the product of tamas or the creatures of the lowest kind as said by Manu.

Fish as Food

Fish has been considered as an item of food either as ordinary victuals or as offerings to the manes and Gods. In the Gautama Dharmasutra, which is perhaps the earliest of all the Sustras (600-400 B.C.), The translated passage runs as follows:

"edh-odaka yavasa-mula-phala-madhv

abhay-abhyudyata-sayy-asan-avasatha

yana-payo-dadhi-dhanna-saphari-priyangu

sran-marga-sakany-apranodyni sarvesam II"

"Fire-wood, water, grass, roots, fruits, honey and fish were recognized food". It was also pointed out a conch, a seat, shelter, a carriage, milk, sour milk, roasted grain, small fish, millet, a garland, *venison*, and vegetables, spontaneously offered by a man of any caste must be refused." The Vasistha Dharmasutra (XIV, 12) has the same passage with certain modifications. The passage includes saphari, meaning thereby small fish. The corresponding text in the Manu Smriti (IV, 250) runs as follows:

"Sayyam grham kusan gadhan

apah pushapam manin dadhi'

dhana matsyam payo mamsam

sakam c-aiva na nirnudet !!".

"A conch, a house, kusa grass, perfumes, water flowers, jewels, sour milk, grain, fish sweet milk, meat and vegetables are not to be rejected it they are offered voluntarily. An almost similar text is found in the Yajnavalkya Smriti (IX, 214).

Cooked fish is to be served in sradh ceremonies. In the Gautama Dharmasutra (XV, 15), it has been claimed that manes are satisfied for three years by eating fish and the flesh of common deer, spotted deer, hares, turtles, boards, and sheep.

In the Apastamba Dharmsutra, reference to a particlar kind of fish, called Satabali has been made,

"khadg opastarane khadga-mamsen-anantyam kalam

tatha satabaler" = matsysya mamsena ...

"If a rhinoceros meat is given to Brahmanas seated on seats covered with the skin of rhinoceros, the Manes are satisfied for a very long time. The same effect is obtained by offering the flesh of the fish called Satabali'. In the Vishnu Dharmasutra, we have:

"dvau masau matsya-mamsean–(LXXX, 2)

"The flesh of fishes (except those species that are forbidden) for two months, which is repeated in the Manusmriti (III, 268). It attaches a greater importance to particular type of fish, called Mahaslka, which when eaten by hermits in the forest serves for an endless time'. In the Yajnavalkya Smriti (X, 258-61) the statement is more elaborated.

Fish Forbidden as Food

In the Gautama Dharamsutra though fishes have been considered as items of food but emphasis has been laid therein that they should not be "misshapen" (XVII, 36). Amongst fishes, the Ceta ought not to be eaten', Nor the snake-headed fish, nor the alligator, not those which live on flesh only, nor those which are misshaped like mermen'. In the Vasistha Dharmasutra (XIV, 41-2), the following types of fishes have been forbidden to be taken as food. Ceta, Gavaya, Sisumara, Nakra, Kulira, misshaped once (vikrta-rupah) and Sarpasirasa. While Mahaslka fish is given a special importance in the Manu Smriti, at the same time, eating of all types of fish has been forbidden therein (V, 14–15) Translated version is given below:

"The Baka, and the Balaka crane, the reven the Khanjaritaka (Wagtail animals) that eat fish, village pigs, and all kind of fishes'. He who eats the flesh of any animal is called the eater of the flesh of that particular creature, he who eats fish is an eater of every kind of flesh, let him therefore, avoid fish."

Religious Injuctions Pertaining to Fish and Fishing

Some religious injunctions pertaining to fish and fishing are found in the Sutras. A snatakabrahmabna is required to wash and touch fire if he eats food containing meat, fish or sesamum (Baudhayana Dharmasutra, Prasna II, Adhyaya III, Kandika 6, Sutra 2).

The course of a year is equivalent to that of a farmer who ploughs with an iron ploughshare in a day. Again, it is stated that the fisherman incurs the same sin as that by a hunter or a fowler (Adhyaya II, VV, 8-9).

Fish Protection

The Visnu Dharmasutra contains some laws which are intended to protect the fish from killing. The killing of fish, even without intention is to be atoned for by fasting and suitable penances. The killer or stealer of fish is required to pay a fine. Thus, it says,

"paksighati matsyghati ca dasa-karsppanan !"

(Adhyaya V, Sutra 53).

"A killer of birds or of fish shall pay ten (10) Karsapanas' (Adyaya V. Sutra 53)". The Manu Smrti lays similar laws for the protection of fish. Fish are not be stolen or killed unnecessarily.

"matsyanam paksinam c=aiva tailasya ea ghrtasya ca !

mamsasya madhunas–c = aiva yac = c= anyat pasusambhavam!!"

The catching of fish is the exclusive avocation of particular communities owning to the social stigma attached to the occupation. The Manu Smrti assigns this task to a particular caste, the Nisada.

"matsy-aghato nisadanam tvastis = tv=avogavasya ca !

mead-andhr-cuncu-madgunam-

aranya pasu–imsanam !!

(Adhyaya X. 48)"

"Killing fish to Nisada, carpenter's work to Avogava; to Medas, Andhras, Cuncus and Madgus the slaughter of wild animals". (Adhyaya X. 48). It has been given both in the Manu Smrti and the

Narada Smrti that fish bones are harmful if swallowed unaware along with the flesh.

"andho matsyam = iv = snati

sa narah kantakaih saha ! yo bhasate = ratha-vaikalyam = apratyaksam sabham gatah !!"

(Manu Smrti, VIII, 95).

That man, who in a court of justice gives an untrue account of transaction or asserts a fact of which he was not an eye-witness, resembles a blind who swallows fish with bones'. (Manu Smrti VIII, 95).

The Narada Smrti draws upon the same metaphor to condemn false testimony given by witness in courts. It says (Chap. II, V.14) that the who having entered the court delivers a strange opinion ignoring the true state of the case resembles the blind man who regardless swallows fish together with bones'.

Identification of Fishes

The fishes referred to in the Vedic literature have been identified by Hora (1953). These are:

Saphari

Its scientific name is *Puntius sophore* (Hamilton), a common and widely distributed carp-minnow found in abundance in shallow ponds, ditches and channels. Gautama Dharmasutra enjoins that this fish should not be refused when offered voluntarily.

Though the fish is of a small size and not of commercial importance, it has been held in great esteem in the Sutras. The Baudhyana Grhyasutra enjoins that on the fifth day after the garbhadhana ceremony, the hosue-holder should stand knee-deep in water in a tank and catch fish with his cloth to feed the cranes. In Bengal, Saphari is still regarded as an auspicious fish. Its name has not changed during all these centuries.

Satabali

Apastamba Dharmasutra enjoins that manes are satisfied for a very long period if Brahmins are fed on this fish. In Baudhayana-Grhyasutra we find that this fish should be offered at the istakahoma. According to Hora (1953), etymological meanings of its name provide a clue for this identification. Sata means hundred and bali means one having spikes (bhalla). If this analysis of the word is accepted, then the fish can be identified as *Mastacembelus armatus* (Lacepede). It is a spiny Eel of great flavour and taste. The fish is abundantly found in shallow water and marshy lands.

Mahasalka

It refers to large scaled and famous hills stream sporting fishes collectively known as Mahseers. The common species among them is *Tor putitora* (Hamilton) which is known as the golden or the Himalayan Mahseer. In Manu Smrti (III, 268) this fish has been referred to in two contexts. The first is that it is eaten by hermits living in forests and secondly, if given to Brahmins, the manes remain satisfied for an endless time. Yajnavalkya Smrti (X, 258-61) is also in agreement with the statement made above. Visnu Smrti (LXXX, 14) considers that this fish, if offered, should always be accepted. In Bengal, the fish is known as Mahasol which is nearer to the original Sanskrit form.

Ceta

In both Apastamba and Vasistha Dharmasutras the eating of this fish has been forbidden. The literal meaning of the word Ceta is servant. Hora (1953) has classified this fish as *Cirrhinus mrigala* (Hamilton) on the following grounds:

1. The eating of Marigal is forbidden in Bengal as it may cause leprosy and other skin diseases.
2. Mrigal is used as a scavenger in polyculture on account of its bottom dwelling habit and feeding on bottom debris. In this way, it acts as a servant so far as the cleanliness of a fish pond is concerned.

Sarpasira

It refers to the snake-headed fishes belonging to the family Ophicephalidae. There are several speices of the genus *Channa* (Ophicephalus) which are commercially important fishes. Since they are carnivorous in their feeding habits, they are included among the forbidden fishes both in the Apastamba and Vasisthan Dharmasutras. It is particularly due to the fact that the head of the fish resembles that of a snake. Moreover, all misshapen fishes are forbidden to the eaten.

Mardura

It means sort and jelly-like fish. The name may refer to Bombay duck. *Harpodon nehereus* (Hamilton) which presents a jelly like in fresh condition. It is not a table fish. It is included among the forbidden fishes in Apastamba Dharamasutra, probably on the ground that the fish presents a ferocious appearance of its jaws. It refers to some species of the genus Tetraodon as this the genus which looks like a man head. It belongs to the group of the poisonus fishes. In Apastamba Dharamasutra, it is included among the category of forbidden fishes.

Sahasradamstra

This fish has been identified as *Eutropiichthys vacha* (Hamilton) on the ground that it possesses a large number of the teeth both in its jaws and palate. It is an excellent table fish. Baudhayana Dharamasutra included among the fishes that would be eaten.

Cilicima

It has been regarded as the famous hilsa fish, belonging to the genus and species of *Hilsa ilisha* (Hamilton) on the ground that clupeid fishes are very common in Indian waters and hilsa can ascend in the river Gangas as high as in Agra and Delhi. Baudhayana Dharamasutra included it among the fishes to be eaten.

Varmi

Its literal meaning is a fish with an armour on its body. A number of catfishers are known to possess armour like bones to protect their body by the reference seems to have been made to *Rita rita* (Hamilton) which is a common food fish of northern India. Baudhayana Dharamasutra include it among the fishes to be eaten.

Brhacchiras

It refer to a fish with a large head. Based on the size and weight of the head to the rest of the body *Catla catla* (Hamilton) seems to represent this fish. Baudhayana Dharamasutra includes it among the fishes to be eaten.

Mahasakari

It is logical to consider maha sakari is mahasaphari which will mean a large Saphari. There are several large species of the genus, *Puntius*. Probably, the reference seems to have been made to Sawarn Puti, *Puntius sarana* (Hamilton) which is a valuable food fish. Baudhayana Dharamasutra include it among the fishes to be eaten.

Rohita

This is the most famous carp fish of India which is still known by the same name in Bengal. It has been scientifically christened as *Labeo rohita* (Hamilton). Baudhayana Dharamasutra includes it among fishes to be eaten. Manu Smrti permits this fish to be eaten after it has been offered to gods as per Yajnavalk Smrti, this fish is fit to be eaten by the Brahmins. Vishnu Dharamasutra also include it among the fishes to be eaten.

Rajiva

It refers to one that moves in a formation. It is *Mugil corsula* (Hamilton) which moves in army-like formation keeping a part of the head out of water. Both Baudhayana Dharamasutra and Vishnu Dharamasutra includes it among fishes to be eaten. In Manu Smrti also, eating of this fish has been permitted on all occasions.

Pathina

It is one of the most famous fishes in ancient Indian literature known as *Wallago attu* (Bloch). Manu Smrti permits this fish to be eaten after it has been offered to the Gods or to the manes thus giving it the same status as to Rohita. Both in Yajnavalka Smrti and Visnu Dharamasutra, it has been considered to be a food fish.

Smhatundra or Simhatundaka

It is fish with the face of a tiger and probably refers to *Bagarius bagarius* (Hamilton). The fish lives in waters of hill streams in association with Mahseer. Manu Smrti permits it to be eaten on all occasions. Yajnavalka Smrti considers it fit to be eaten by Brahmins Vishnu Dharamasutra also considers it as a food fish.

Sasalka

It refers to fishes with scales as a whole, mostly carps. They are permitted to be eaten on all occasions in Manu Smrti.

Sakula

This refers to the sal or sol fish of Bengal known as *Ophicephalus striatus* (Bloch). It is a snake-headed fish belonging to the family Ophicephalidae. Eating of the fish has been forbidden in both– Apastamba and Vasistha Dharamasutras but permitted in the Visnu Dharamasutra.

It can be concluded that during the period 600 B.C. to 200 A.D., although fish was considered to be a valuable article of food among the Hindus, but at the same time certain species were forbidden to be eaten for one reason or the other. Even among those considered suitable for eating, there was a regular gradation on quality or value. For instance, Mahasalka or *Tor putitora* was assigned the highest rank, then came Rohita (*Labeo rohita*), Pathina (Wallago attu), Simhatunda (*Bagarius bagarius*), Satabali (*Mastacembelus armatus*), Saphri (*Puntius sophore*) and so on. Again certain aquatic animals and fishes were forbidden to be eaten. The Smrtis contain contradictory statements about the use of fish as food. This reflects that the social, religious and political structures were being changed. The result was that the taking of fish or animal flesh became a taboc afterwards.

Fish and Fisheries in the Ramayana and Ancient India (200 B.C.–700 A.D.)

The names of three fishes have been frequently referred to in the Ramayana. These are:

Cakratunda or *Garra mullya* (Sykes)

The fishes of the genus *Garra* are commonly found in rocky streams and clear lakes. They possess a suctorial disk on the ventral surface of their head. The suctorial disk works as an adhensive organ by means of which they fasten themselves to rocks and stones in swift currents. Cakratunda is thus an appropriate name for the fish with a suctorial mouth. Even in calm and clear waters of lakes, the fish adheres to objects for rasping off algae salimes which forms their food. The fish does not grow to large size and contains much fat. It is greatly relished by the hill people.

Nalamina or Mastacembelus armatus (Lecepede)

It is the spiny-eel of fresh waters of India. It grows 2 to 3 feet in length. The fish is compressed from side to side like a reed. It has an elongated and tapering snout. Since the fish is like a reed in shape, the name Nalamina is very appropriate for it.

Rohita or Labeo fimbriatus (Cuvier & Valenciennes)

In Northern India, one is familier with Rohita which is *Labeo rohita* (Hamilton) commonly known as Rohi fish. Rohi does not extend beyond the Mahanadi and the Godavari river systems in the South. It is replaced by another species of the genus, *L. fimbriatus*. The fish is Rohita in Sanskrit literature. It grows to about 3 feet in length. It is highly esteemed as food fish in Southern India.

In the north-western recessional of Ramayana, instead of Cakratunda and Nalmina, Sakula nad Pathina are mentioned. Sakula of the Ramayana or Saula of Bengal is snake-headed fish, *Ophicephalus striatus* (Bloch). This fish possesses an air breathing organ and can thus live out of water for a considerably long interval of time. It is known to undertake over land journeys from one pond to another during rainy season. The arrangement of the scales on the head looks-like that of snakes. It is usually called a snake-headed fish. The Sanskrit name Sakula is thus appropriate for the fish. The fish grows to a size of 2 to 3 feet and in Peninsular India it is greatly prized as a food fish.

Pathiana of the Ramayana or Boal is a cat-fish known as *Wallago attu* (Bloch). It is a predatory fish, with a large gape of mouth and jaws studded with numerous sharp and backwardly directed teeth for holding prey. It grows to a size of nearly 3-5 feet.

There are various references to fish and other aquatic animals in general. These include the following:

1. Aranyakanda, 17, 24–Agastya, who was entertained by the gods with Indra for the purpose of killing the demons, drank the ocean full of timi and makra.
2. Aranyakanda, 45, 13–The pure ones, coming in contact with sins which are not committed by themselves become annihilated for no fault of theirs, just as fish become annihilated in a lake infested by serpents.
3. Aranyakanda, 58, 38–Lotus plants, as if with blossoms torn off and fish running to and fro, seem to sympathize with Sita as for their friend.

4. Aranyakanda, 60, 12–When Sita was being carried away, the ocean with its furious fishes and serpents became very boisterous with waves and sprays.

It is evident from the above mentioned references that the Indian waters harboured a big population of fishes during the Ramayana period. References to other equate animals need some clarification in regard to their identification.

Timi is usually translated as 'Whale' though true whales are not very abundant in the Indian seas. Probably reference is made to sharks, which are fairly abundant in the seas.

Nakra is usually translated as alligator bu true alligator are not found in India. In India, both long-snouted and short-snouted crocodiles are found in the rivers. The short snouted crocodiles (*Gavialis gangeticus*) may mean either in saw-fish or the sword-fish of Indian seas. Since the refernce to this animal is found as early as Taittiriya Samhita, it may mean a fish with pointed snout.

Makura, has been referred to in the Yajurveda Samhita as the objects of sacrifice in Aswamedha. It has been rightly understood to represent the short-snouted crocodile. In later literatures the animal has become the vehicle of Ganga and Varuna, the lord of waters.

Description of the Pampa Lake in the Ramayana

The Pampa lake is described as a fine and transquil sheet of water blooming with white and red lotuses. It was devoid of gravel and coarse aquatic plants. Swans, cranes, ospreys etc., sporting in the water are said to have filled the atmosphere with musical nots. The water is said to have sweet smell like that of lotus. The bank are stated to be full of trees such as tilaka, ashoka, pannaga, vakula etc. crested with blossoming flowers. The asylums of Savari and Matanga were situated on the banks of the lake. A portion of the lake was also known as Matangsara. Presumably on the western bank of the lake, stood the forbidding. Rsyamukha mountain, aborned with many a blossoming tree and guarded on all sides by the little serpents. The monkey kings, Surgiva, lived in this mountain.

According to Chacko (1952), the Pampa Sarovar is a rectangular tank measuring 240 feet x 155 feet with an average depth of 4 feet. There is an unprotected channel that leads from the Tungabhadra river to this tank. There is a small circular pond known as Mansarovar which is about 15 yards away from the Pampa Sarovar. It is devoid of outlet and intlet.

The changes that have taken place in the fish fauna of the lake are obviously the results of a lacustrine condition changing into a marshy environment. The fauna obviously changes due to flood in the Tungabhadra River. Fishes like *Labeo fimbriatus* and *Barbus* (*Tor*) *khundree* repopulated Pampa Sarovara. When the tanks dry up, the water-breathing species die out. Genera like *Mastacembelus* and *Ophicephalus* survive as these are known to tolerate marshy conditions. Although *Garra mullya*, *Labeo fimbriatus* and *Wallago attu* was quite common in the Tungabhadra river, their presence in the present day Pampa Sarovara is lacking due to lack of clear water, rocky substratum, occasional desiccation etc.

"Lakshmana was advised to use an arrow (shot from a bow) for catching fish (iii 73) in the crystal clear water of the Pampa Lake and in the marshy areas around, full of reed. The second method of catching fish is referred to by means of a baited hook or angle. In Aranyakanda (51, 27) Jatau is said to have told Ravana, when he was running away with Sita, to the following effect:

"You have come in contact with a deadly snake where from you will not get any relief, just as a fish cannot escape and is ultimately killed when it swallows an angle baited with meat".

It would appear from the above that the practice of catching fish by angling was fairly widely practised in the Ramayana period.

"At the Lake Pampa, Lakshmana was advised to have the scales cleared and the fishes roasted in an iron pan over the fire (iii 73, 15).

"पम्पायामिषुभिर्मत्सयांस्तथ्व राम वरान् दतान् ।
निरस्त्वकपक्षानयरस्त प्तानकृशानेककण्टकान ।।"

In Aranyakanda (76, 24)

स्थानं प्रपचतां तथ दृष्टा मौयारत्तण्डतान् ।
थ्पम्पलौलवणाभ्यां च मत्स्यान् संपादपिष्यथः ।।

Rama and Lakshmana were advised to cook rice and fish with salt and red pepper on reaching the asrama on the west bank of the Pampa Lake. Thus, roasting fish on fire and the use of salt and red pepper appear to be very ancient traditional practice in India. The Mahabharta (400 B.C.) is full of references to a large variety of animals many of which are common to Ramayana and Vedic literatures. Vedavyas, the writer of Mahabharta, is a son of Matsyagandha and sage Parasar. Arjuna of Mahabharta married Drupadi by shooting the target of a fish eye. His daughter in law Uttara, wife of Abhimanyu, is said to have a fish origin. Sage Sauvari was attracted to a family life by seeing a fish moving happily with its off-springs.

Animal Systematics in Hindu, Buddhist and Jaina Culture (600 B.C.–500 A.D.)

There has been a lot of references regarding animal sacrifices in the Vedas but unfortunately the zoological knowledge of the Vedic period has not yet been fully enunciated. The Chandogya Upanishadd has classified animals on the basis of their ovum or seed (bija) into three groups:

1. Born of eggs (ANDAJA)
2. Born of fully young ones or viviparous (JIVAJA)
3. Born as plant-like organism (UDVIJJA)

Further the animals have been classified into eight classes on the basis of their feeding habits. The animals living in water logged areas, marshy lands or even grazing on river bank were called Anupa. The class Anupa appears to represent an important class as it also included the group "Matsya" which according to Rao (1956-57) represented both fresh water sea water fishes.

Chakra's Classification of Animals

The Chakra gives classification of animals based on therapeutic values of fish. The dietic value of fishof an animal was conceived to depend mainly on its habitat and mode of life it lead. Dietary animals were accordingly divided into eight classes including fishes in the class Anupa. Chakra further classified animals into four primary divisions which are based on their origins. Thus we have:

Jarayuja

Born from uterus (Placentals)

Andaja

Born of an ovum or egg (Fishes, Reptiles, Birds etc.)

Svedaja or Usmaja

Born of moisture and heat spontaneously or sexually (Worms, mosquitoes)

Udvija

Born sexuallyor asexually both

Classification of Animals of Susruta

"The Susrita classification of animals is based on their origins. The animals were divided into four groups.

1. Jarayuja or (viviparous)–*e.g.*, mammals
2. Andaja or (oviparous)–*e.g.*, fishes, birds, reptiles
3. Svedaja (born out of heat and moisture)–*e.g.*, ants, worms etc.

The next gives two other system of classifications. These are based on the habits and anatomical features of the animals. It takes into account only those creatures whose flesh is edible. Thus we have six categories of animals:

1. Jalesaya (aquatic)
2. Anupa (frequenting marshy lands)
3. Gramya (domesticated)
4. Kravyabhuja (carnivorous)
5. Ekasapha (uni-hoofed)
6. Jangala (dwelling in dry areas)

The second system of classification is based on habit and habitats of the animals. These are (1) Jangala (animal that live in jungle, dry or hily lands) and (2) Anupa-aquatic or amphibious animals (animals living in marshy or water logged areas).

The class Anupa is further divided into five sub-classes. The fishes were given place in sub-class, 'Matsya' which included both fresh water and sea water fishes. About more than twelve inland of fresh water have been described in this text.

Classification of Animals of Prasastapada

The animals were classified according to whether they were born sexually or asexually. The asexually born were considered to have born from the eggs. Sexually born were further categorized according to whether they possessed bones and blood and whether they could be crushed easily or not.

Umasvati Classification

Umasvati in his work Jattavarthadhigama evolved another method by which animals were classified into four categories based on the number of senses possessed and used by them. The fishes were assigned of the fourth category in which animals with five well developed senses of touch, tase, smell, sight and hearing were placed. The fourth category was further subdivided according to the mode of reproduction. Fishes were placed in the sub-group "ANDAJA".

Susruta Contribution to Ichthyology (400 B.C.)

Fish as food

Sasruta not only enumerated a number of fish species that have food value but also gave the relative medicinal properties possessed by different fishes. Hora (1955) stated that fresh water fishes are more nutritious than the marine fishes. According to him, the non-protein nitrogenous components such as trimethylamine oxide and urea which are present in marine fishes, putrefy more readily and thereby reduce their palatability and sale value. The fresh water fishes on the other hand do not contain these components and can stay fresh for a relatively longer period. During ancient time, when refrigeration was not known, the relatively high merit of fresh water fish as an item of diet in India could thus be readily understood.

Besides enumerations of a large number of fresh water and marine fish species in Susruta Samhita, there is a significant reference to correlation between the form and locomotion of fishes and to the influence of diverse habitats in moulding their body forms. A passage in Susruta runs as follows:

नदियां गुरवो मध्ये यस्मात् पुच्छास्यचारिणः ॥

सरलङ्गगजानान्तु विशेषणं शिरो लघु ।

यदरगोचरा यस्मात्तस्मादुत्सोदपानजाः ॥

किचन्गुक्त्चा शिरोदेशसत्यर्थ कि गुरवस्तु ते ।

थांधसादुगुरवो ज्ञेया मत्सयाः सरसिजाः स्मृताः ॥

डरोविचरणात्तेहाां पूर्वमगं लघु स्मृतम् ।

खड्त्यानूपो महाभिव्यारदघातयः दिमासंव ॥

Kaviraj Kunja Lal Bhisagratna (An English translation of the Susruta Samhita, I, P.493. Calcutta, 1907) translates this passage as follows: River-fish are heavy at the middle, owing to the fact of their moving about with the help of their head and tail, while those which are cultured in tanks and ponds (Sarah and Tadaga) are specially light about their heads. Fish, which are found in hill streams or mountains, are extremely heavy about the parts a little below the region of their head, on account of their being confined within narrow limits and the consequent absence of a lengthy sweep. Fish reared in large tanks (Sarasi) are lighter in the foreparts of their body and heavy in their lower parts, as they put their entire pressure upon their breast at the time of swimming. However, a free translation of the passage may be given as follows:

The river fish are bulky in the middle because they move with their head and tail; the lake and tank fish are similar to the above but are characterized by a relatively smaller head; are extremely deep behind the head; the fishes of the torrents are traditionally well known by the possession of two characteristic; the greatly flattened body on account of their habit of drawing with the chest, and relatively reduced anterior part of the body.

According to Susruta, the fresh water fishes are grouped into four ecological 'formations'. These are:

1. River fishes, bulky in the middle (wedge-shaped)
2. Lake and tank fishes, bulky in the middle but with relatively smaller heads.
3. Spring and pool fishes, lengthening of the vertical axis.
4. Torrential fishes, greatly flattened and depressed form.

Thus, we see that even reasons are assigned for the correlation between the different forms and the corresponding habitats. For instance, their river fishes are wedge-shaped because they move fastly with their head and tail. The vertical axis of the spring and pond fishes is elongated because they do not move about much. The fishes of torrential streams are greatly flattened because they have to crawl on rocks with their chests.

It is remarkable that the ancient Hindus had a fair knowledge of the involvement of both the anterior and posterior parts of the body in the movement of fishes. This was discovered by Breeder in 1926 and Gray in 1933. Gray (1933) demonstrated that in the locomotion of fishes the initial movement starts from the anterior region of the body and that the tail also exerts a fraction of the forward propulsive thrust.

The ancient Hindus were also familiar with our present concept of ecological segregation and fish locomotion. They also know the laws governing the correlation between body-forms and habits and habits and habitats.

The exact age of Susruta's monumental work is very difficult to determine. However, Bhisagranta discussed the age and personality of Susruta and came to the tentative conclusion that 'At all evetns, Nagarjuna who redated the Susruta Samhita lived abou the later part of the fourth century before the Christian era ....'. Thus there seems no doubt that several centuries before the Chiristian era, Hindus had a fair idea of correlation between the body form and locomotion in fishes. The influence of diverse habitats in bringing about a change in the body form has been nicely illustrated by Hora (1935).

Fishereis Science in the Budhist Period and the Magdhan Empire (600 B.C.–232 B.C.)

Kautilya Arthasastra is the earliest known work (400 B.C.) of the ancient Hindus. The life of the people and their agriculture including fisheries science during the Budhist period and in the Magdhan empire are well documented in it. Then we have Jataka stories which tell us about the previous incarnations of Lord Budha, the sculptures of the railings and gateways of the stupas of Bharhut and Sanchi (200 B.C.).

The Nanda dynasty was over thrown by Chandragupta Maurya in 325 B.C. whose chief adviser was Kautilya, also known as Chanakya or Vishnugupta. He was the author of the Arthasastra. Hora (1948) consulted Shamasastr's English translation of the Kautilya Arthasastra (2nd edition, Mysore, 1923) and made the following references from this translation.

Book IV, Chapter III, deals with remedies against national calamities under famines, Kautilya inter alia advise the king to remove himself with his subjects to sea shores or to the banks of rivers or lakes. He may cause his subject to grow grains, vegetables, roots, and fruits wherever water is available. He may, by hunting and fishing on a large scale, provide the people with wild beasts, birds, elephants tigers or fish'. (p.254).

Book XIII, Chapter V, deals with the operation of a siege, Kautilya advise the reduction of the enemy before the commencement of the siege and alongwith a large number of the other measures suggests:

'A splinter of fire kept in the body of dried fish may be caused to be carried off by a monkey, or a crow, or any other bird (to the thatched roofs of the houses)' (p.468).

The passage is significant for it shows that in the ancient days the invading armies carried a supply of dried fish with them. As such, the art of processing fish must have known to those people. It can be presumed that when a splinter of fire is kept in the body of dried fish. It is not likely to go out since there is always a certain amount of body oil that will come out from dried fish when heated and this will keep the splinter alive. The monekys, crows and other birds, as usual, are fond of fish.

The king shall exercise his right of ownership (svamyam) with regard to fishing, ferrying and trading in vegetables (heritapanya), in reservoirs of lakes (Setushu).' The reservoirs referred to in this passage appear to be irrigation tanks. The reservoirs or tanks in those early days were used for fish culture enterprise (Book II Chapter I page, 57).

Book II deals with the 'Regulation of toll-dues'. As contained in Chapter XXII, (p.135), imported commodities were charged one - fifth of their value as toll. Incase of perishable articles such as flowers, fruits, vegetables (saka) roots (mula), seeds, dried fish and dried meat, the toll was one-sixth of the value of the articles. With regard to conch shells the fixation of the amount of the toll was left to the judgement of experts. The noteworthy point is that dried fish must have been a regular trade commodity in those days and therefore, the methods of processing fish must have been known to the people.

Among the directions given to 'The Superitnendent of Agriculture' (Book II, Chapter XXIV, p.141), it is stated that 'sprouts of seeds, when grown, are to be manured with a fresh haul of minute fishes and irrigated with the milk of snuhi (*Euphorbia antiquorum*)'. It is interesting to note that small fishes of little economic importance as table fish were used as manures and that the high value of fish manures was well recognized by the ancient Hindus.

The Superintendent of slaughter hosues (Book II, Chapter XXVI, pp.147–148) is enjoined to enforce several regulations concerning fish and other animals. In regard to fish:

1. When a person entraps, kills or molests fish which are declared to be under state protection or which live in forests under State protection (abhayaranya) he shall be punished with the highest punishment.
2. When a person entraps, kills, or molests fish that do not prey upon animals, he shall be fined $26^{3/4}$ Panas.
3. Of fish that have been captured, the Superintendent shall take one-tenth or even more than one-tenth as toll.
4. Fish in tanks, lakes, channels and rivers shall be protected from all kinds of molestation.
5. Fish living in forests under State protection shall, if they become of vicious nature, be entrapped and killed outside the forest preserve. Among the instructions given to 'The Superintendent of Ships' (Book II, Chapter XXVII, pp.152–153), the following guidelines are provided:
   (*a*) Fishermen shall give–sixth of their haul as fees for their fishing licence (naukahatakam).
   (*b*) Fishermen shall be exempted from payment of fees for fording or crossing rivers at any time and place.

It remains for the Fishery Department to examine the above mentioned regulations and to compare them with the existing rules and regulations in their respective areas. As contained in the Kautilya's Arthasastra, the fishing fees charged by Zamindars and other privagte agencies of fisheries should not exceed one-sixth of the haul. Usually these days it is of one-half or even up to two thirds of their haul. The fishermen are subjected to such an amount of rent or other imposts that they can hardly carry on their profesion.

In Book I, Chapter XIII, p.24,–we get the following reference:

'People suffering fish swallowing smaller ones (Matsyanyayabhibhutah).

In Book II, Chapter IX-p.77, it is referred, just a fish moving under water cannot possibly be found out either as drinking or not drinking water, so government servants employed in the government work cannot be found out (while) taking money (bride).

Kautilya deals with secret means to injure an enemy and suggests the following mixture and treatment for rendering fishes poisonous (Book XIV, Chapter I, pp.478–479).

'The mixture prepared from the flowers of Bhallataka (*Semecarpus anacardium*), yatudhana, dhamargava (*Archyranthes aspera*) and bana (sal tree), mixed with the powder of ela (large cardamom), kakshi (red aluminous earth), guggulu (ibdellium), and halhala (a kind of poison), together with the blood of a goat and a man, causing biting madness'.

'When half a dharana of this mixture, together with flour and oil-cakes, is thrown into water of a reservoir, measuring a hundred bows in length, it vitiates the whole mass of water; all the fish swallowing or touching this mixture become poisonous; and whoever drinks or touches this water will be poisoned'.

There seems a great scientific knowledge in the practice mentioned above. In the first place, a distinction is made between poisoining and rendering fishes poisonous. When this treatment is carried out in a water body, the fishes do not die but their flesh becomes poisonous so that whoever eats it, is poisoned.

It is well known that when carps are fed on silkworm pupae, their flesh becomes poisonous. It is the usual practice to condition such fishes for three or four days before they ar sent to the market for sale. It is also known that certain fishes becomes poisonous during certain periods of the year. This is particularly due to the fact that during these periods fishes feed on certain types of poisonous algae. The fish culturists know that the quality of the food determines the taste and flavour of the fish.

People in the hill region use several kinds of herbs, fruits, barks of trees etc. for poisoning or stupefying fishes for their capture. The object is not to injure the fishes in any way but to render their flesh poisonous for the enemy. It is clear from Kautilya's account that fish was cultured in reservoirs and that it used to be captured for human consumption. Moreover, fresh fish used to be a regular item of ration for the army.

Fish was also used as one of the ingredients alongwith other mixtures to cause injury to an enemy (P.476).

Though there is a well known saying "to drink like a fish" but Kautilya was right in saying that it is not possible to determine whether a fish drinks water or not as we do. His analogy of government servants taking bribes is also appropriate.

Fish and Fisheries in the Jataka Tales (380 B.C.–500 A.D.)

The word Jataka, meaning birth, has a special Buddhist connotation. The Buddhist believe that one cannot attain Buddhahood (*i.e.*, the state of supreme wisdom) in one birth alone, but must pass through the numerable existences as a Bodhisattva. In each such existence the Bodhisattva, through a life of piety, develops his moral, intellectual and spiritual powers until he becomes a Buddha. When he reaches this supreme state, he acquires knowledge of former lives. These stories of his former lives are collectively known as the Jatakas. Nearly 547 storeis to the categories of folklores, anecdotes, fables and fairy tales, as a collection, are of considerable value since they contain a valuable record of the life and thought of the people during that period.

Hora and Saraswati (1955) commented on the references made regarding fish and fisheries in the extensive Jataka literature comprising 547 storeis. According to them the varieties of fish metioned in the Jatakas and their indentifications are as follows:

Ananda-machchha

This is referred to in Jataka 32 (Text, Vol. I, p.207; Trans; Vol. I, p.83) where it is stated that fishes chose Ananda as their king, just as quadrupeds chose lion as their king.

The translator adds 'monster-fish', but there is no warrant for this qualification in the text. Rhys Davids and Steed in their Pali-English Dictionary (P.T.S.) translate the word Ananda as 'Leviathan'.

The largest known fish is the Whale Shark, *Rhincodon typics,* which grows to a large size and can easily swallow sharks of 25 feet or so. There is a representation of this fish among Barhut Sculpture. Probably this fish in Sanskrit literature is referred to as Timingila-gila as the name Ananda has no corresponding name in the Sanskrit texts. The translation as 'monster-fish' in the Sanskrit text makes it clear that Ananda refers to Whale Shark, *Rhincodon typicus*.

Kanamahamaccha

This is referred to in Jataka 38 (baka-Jataka, Text, Vol. I, p.222: Trans., Vol. I, pp.96–97). The story relates to a crane deceiving some fish of a dyring pond by promising to carry them to a big pond but eating them on the way. The first of the fishes thus deceived was kanamahamaccha (one-eyed; large-sized fish), whom the other fishes deputed to go first in the hope that if the crane were to play any trick the big fish would be a match for it 'whether afloat or ashore'.

Accoding to Hora and Saraswati (1955), the large size of the fish suggests that, apart from the freshwater shark known as the Boal (*Wallago attu*), it may be Catla (*Catla catla*) or the Rohu are frequently represented. The reference to blindness in one eye possibly symbolizes the ease with which the fishes were deceived. However, rare examples of one-eyed carps are known: one of which has been described by Nair (1940).

Mahamukhamaccha

This fish is referred to in Jataka 288 (Text, Vol. II, pp. 423–424; Trans. Vol. II, pp.288–289) as to have swallowed and protected a packet of money which was dropped in the river Ganges. Later, when the fish was caught, the money was recovered intact from it.

Mahamukhamaccha means a fish with a large mouth. In river Ganges, two cat fishes are known which posses large mouths and are voracious in habits-Boal (*Wallago attu*) and the Bachha (*Eutropiichthys vacha*). The boal grows to about 5 feet in length and could easily swallow a packet of money and retain it in large and bag-like stomach. Accordingly, Mahamukhamaccha is *Wallago attu* (Bloch).

Rohita

This is the most celebrated freshwater carp of northern India. It is known as *Labeo rohita* (Hamilton). References to this fish in the Jataka regarding its food value and medicinal properties are many. For example, in Jataka 293 (Text, Vol. II, p.433; Trans; Vol. P.295), it is stated that abdominal pain (udaravaṭa) subsided soon after a meal of rice, fresh ghee and Rohita was taken …. In Jataka 316 (Text, Vol.III, pp.51; Trans; Vol. III, pp.35–36), the abundance of Rohita in the Ganges is indicated by the fact that a fisherman catches seven in a morning. The taste for this is also brought out in the story and it is shown that it is a suitable offering for Brahmanas. In Jataka 400 (Text, Vol. III, p.333; Trans. Vol. III, pp.205–206) it has been recorded that jackals and others are fond of Rohita. This story is depicted in Bharhut sculptures also. In Jataka 453 (Text, Vol. IV, pp. 72–73; Trans. Vol. IV, pp. 46–47) the sight of a Rohita fish first thing in the morning is considered a good omen for the day.

Velamaccha

This is referred to in Jataka 402 (Text, Vol. III, p. 345; Trans., Vol. III, p. 212). It is known as a ravenous, marine fish. The Jataka reference is made to a human being seized and dragged away by this fish. Evidently the reference here is to sharks, which abound in Indian seas and are greatly dereaded by coastal fishermen. Probably, the reference is made to the man-eating shark, *Carcharias gangeticus*.

Ksura-nasika

In Jataka No.463 (Text, Vol. IV, p. 139; Trans., Vol. IV, pp. 88–89), reference is made to species of marine fish with a body like that of men and with a sharp razor-like snout. They are stated to dive in and out of the water in open sea. A shark fish agreeing with this brief description would be the Saw-fish, *Pristis, Lauspidatus*.

Suvannavanna-maccha

In Jataka 491 (Text, Vol. IV, p. 335; Trans; Vol. IV, p. 212) reference is made to golden-coloured fishes as auspicious. The Saphari, *Puntius sophore* (Hamilton) is probably referred to in this Jataka as the fish is golden on the gill-covers. It is still regarded as auspicious in Bengal.

Kumbhila

In Jataka 529 (Text, Vol. V, p. 255; Trans; Vol. V, p. 131) it has been referred to as a kind of marine fish. Kumbhila means pitcher-like so the reference seems to be to ball or ballon-fishes belonging to the family Tetraodontide.

Makara

This is referred to in Jataka 529 alongwith the Kumbhila. Makara is a well-known Sanskrit name for the crocodile (*Crocodilus porosus*). Crocodiles rarely enter the sea but in those days the crocodile, ebing aquatic, was often errouneously classed with fishes.

Susu

This referred to in Jataka 529 along with the Kumbhila and Makara. The Susu is still used for the Indian freshwater porpoise. *Platinist gangetica.* These fresh water mammals are still popularly regarded as fishes.

Pathina

In Jataka 545 reference to a magic jewel is made in which rivers were visible and Pathina, *Wallago attu* (Bloch) could be seen in these rivers. This fish occupies a place of honour in Smriti literature and is referred to as a strong fish in Jataka 451. In the magic jewel, the following fishes could be seen along with Pathina.

Pavusa

This fish is referred to in Jataka 545 and 451. It can be identified as Pabda, *Callichrous pabda* (Hamilton), a delicious and small-sized cat fish.

Valaja

This fish is referred to in Jataka 451 and 545. Vala maccha also occurs in Jataka No.402. As vala means snake the reference seems to have been made to the spiny eel, *Mastacombelus armatus* (Lacepepede) which grows to a fairly large size and is a good edible fish.

Munja-rohita

This is referred to in Jataka 451 and 545. The significance of Munja when prefixed to Rohita is not understood. Possibly the name Munja-rohita has been referred to Rohita (*Labeo rohita*) feeding on Munja a type of weed, which provides the fibres form of which ropes are made in northern India, It may acquire a different taste and flavour.

Prthuloma

This fish is referred to in Jataka 545. Prthuroma has been synonymized as fish in general though it means fishes with big scales. It is probable that Prthuloma was used for all varieties of scaly fishes.

Besides the above specific names used for certain varieties of fishes, Mahamaccha has been used to denote a large fish and Khuddaka-maccha for a small fish in general.

The Habits and Instincts of Fishes in Jataka Tales

Several Jataka tales refer to the intelligence of fish. In Jataka 311, we have the story of Sakka taking the form of a fish and feigning dead to test the vow of a crance pledged not to kill living creatures. Again, we have the story of a fish, called Thoughful (Mitacinti), splashing in water upstream and down-stream, to give the impression to the fishermen that the net had been broken through with the purpose of releasing its two comrades traped in the net. This fish is said to have already apprehended such a danger from nearness to human habitation and to have advised the two comrades to leave the locality forthwith. Fish are also known to splash in water to frighten away cranes. They move in shoals and act in unison to destroy a common enemy.

Another story states that the fish know by instincts when the year will be rainy and when there will be drought. They accordingly move into deep waters to avoid drought. Those which fall to do so in good time conceal themselves in the mud to avoid being caught. It is also said that fish migrate when their numbers are reduced as they feel unprotected.

Two of the Jataka tales refer to fish having a natural attraction for music. One story says that fish in a lake followed the king of Banaras because they were attracted by the music of his retinue. Then the king instructed his attendants to sound a drum to call the fish to be fed. In another story, ship was wrecked at sea through the singing of a minstrel accompanied on the lute. The music attracted a swam of fish among which there was a sea-monster. The monster leapt on to the vessel and broke it.

Devotion to their mates is another characteristic of certain fishes known to early observes. Two of the Jataka stories relate how a trapped fish in the face of death laments the fact that its mate might regad. It is faithless.

Voraciousness and immoderation of fish are frequently mentioned and people are warned not to have a fish-like tongue. Fish generally live on rock-sevala (Vallisneria) but rice is also one of their favourite foods. Some species are cannibals. As such it is the usual practice for bigger fish to swallow smaller ones. Anandamachchha, a monster fish, was appointed the king of the fishes. By accident it tasted a fish and enjoyed it so much that feasting on its subjects became a habit, which almost caused his own destruction. The fishes migrated and Ananda-machchha went in search of its favourite food. One day it saw the departed fish hiding in a certain mountain which he encircled to prevent their escape. Mistaking its own tail for a fish 'Anandamachchha devoured it with a crushing sound, suffering thereby a lot of bodily pain.

Methods of Catching Fish

Different methods of catching fish, nets and basket traps have been mentioned in the Jatakas.

Fish as food

Fish-eating habit was widelhy prevalent in the days of Jataka tales. Even ascentic enjoyed fish dishes. In one of the tales, it has been claimed that no sin is attached to a holy man if he eats fish, the sinner being the one who kills the fish. Ajivikas, a religious order of naked ascentic, are also said to have a fish diet.

Women are regarded as having a particular yearning for fish. In the story, a thief gives woman 'fish, flesh and strong drink' with the intention of killing her and stealing her ornaments. In another story, the Bodhisativa advises a young man to return to the hermitage when he is tired by the demands of his beloved for fish and rich food.

One of the usual modes of preparing fish for eating was to roast it in embers. There are evidences of fish being dressed and richly cooked. Fish and mean preparations were usually greatly relished with strong drinks. One of the tales refers particularly to the satisfaction obtained by having a dainty dish of stinking dried fish with spirits.

Fish Omens

Fish were regarded as good omens. In one of the Jataka stories it is said that it is auspicious to rise from bed and to see, among other things, a Rohita.

Fish and Human Settlements

The availability of fish seems to have determined the choice of sites for human habitation. This is particularly because the fish constituted one of the principal items of food for the people. In Jataka 547, a certain king is advised to dwell on the river Ketumati because it was full of fish.

Fishing Villages

Jatak 41 speaks of a fishing village in the kingdom of Kosala as consisting of 1,000 families.

Sales of Fish

Fish, as already noted, was one of the principal items of food, and Jataka 288 contains an account of fish being sold in the streets.

Fish and Fisheries on Ashoka Pillar Edict V (246 BC)

Ashoka succeeded Bindusara on Mauryan Throne in about 246 B.C.

Kalinga (Orissa) war, he became a Budhist and abjured violence. In his inscriptions the king is styled as Devanampiya Priyadarsi raja (laja) *i.e.*, 'King Priyadarsi, the beloved of the gods'. Inscriptions of Ashoka have been found engraved on rocks, separate stone blocks and stone pillars. These have been designated as dhamnalipis, translated as: edicts of the law of piety (morality)". Three categories of inscriptions on rocks are recognized. These are:

1. Fourteen rock edicts in seven or eight (if the Sopara version of the edict VIII is taken to imply the existence of thirteen others at the place) recensions.
2. Two rock edicts separately incised at Dhauli and Jaugada each in two recensions.
3. One minor rock edict in ten recensions.

On the inscriptions of stone block, the Royal Asiatic Society of Bengal has the Calcutta-Bairat edict enumerating the sacred text of the doctrine. The pillar inscriptions fall into two groups:

1. Seven pillar edcits, the first six in six recensions with the seventh on the Delhi-Topra Pillar.
2. Minor pillar edicts.
   (*a*) One schism edict in three recensions.
   (*b*) Queen's edict in one recensions

These inscriptions are of paramount importance for a study of the successive stages of the working of the mind of Ashoka who has been regarded as 'one of the greatest personalities of world history.'

Ashoka 's Pillar Edict V

Ashoka's pillar edict V, which shows his Dhammaniyama or regulation of poety, has been found without any textual variation from six places, namely, Topra (180 miles from Delhi on the direct line between Ambala and Sirsava); Mirat, U.P.; Radhia (=Lauriya), Champaran Dist; North Bihar; Rampurva 32 miles north-west of Bettiah, Champaran Dist; North Bihar and Kosam, Allahabad, Distt., U.P. It deals with regulations for the protection of many varieties of animals. The following is its free translation after Hultzch (1925).

King Devanampriya Priyadarsin speaks thus:

The following animals were declared by me inviolable, viz; parrots, mainas, the aruna, ruddy geese, wild geese, the nandimukh, the gelata, bats, queen-ants, tortoises and porcupines, squirrels, the srimra bulls set at liberally, iguanas, the rhinoceros, white doves, domestic doves, (and) all the quardrupeds which are neither useful nor edible.

Those (she-goats), ewes, and sows (which are) either with young or in milk, are inviolable, and also those (of their) young ones (which are) less than six months old. Cocks must not be caponed.

Husks containing living animals must not be burnt.

Forests must not be burnt either uselessly or in order to destroy (living beings).

Living animals must not be fed with other living animals. Fish are inviolable, and must sold, on the three Chaturmasis (and) on the Tishya full-moon during three days (viz) the fourteenth, the fifteenth, and the first (tithi), and invariably in every fast-day. During these same days also no other classes of animals which are in the elephant part and in preserved of the fishermen must be killed.

On the eighth (tithi) or (every) fortnight, on the fourteenth, in the fifteenth, on Tishya, on Purnarvasu, on the three Chaturmasis, (and) on festivals, bulls must not be castrated (and) he–goats, rams, boars, and whatever other (animals) are castrated (otherwise), must not be castrated (then).

On Tishya, on Punarvasu, on the Chaturmasis, (and) during the fortnight of (every) Chaturmasis, horses (and) bullocks must not be branded.

Until (I had been) anointed twenty-six years, in this period the release of prisoners was ordered by me twenty-five (times).

Five varieties of fish or fish like animals are included in this edict. These five species of fish in this pillar edict were discussed and identified by Hora (1950), as follows:

Anathikamachhe

Reference is probably made to 'boneless fishes' or sharks (Elasmobranchii) which sometimes ascend rivers for considerable distances into fresh waters.

Vedaveyake

It may be identified as something eleding grasp or an eel fish.

Gamgapuputake

It has been identified as *Platynista gangetica* (Indian fresh water porpoise), fishlike creature with a humpy body.

Samkujamachhe

the literal meaning of the name would be 'contracting fish'. Thus it has been identified as a skate or ray. These are the fishes which moved by contracting and expanding their bodies.

Kaphatasaykae

It has been identified as globe fish. The globe fishes when frightened, gulp in air and are able to inflate themselves as balloons. They then turn upside down and float helplessly, feigning death.

The above mentioned five varieties of fishes (animals) were declared inviolable by Ashoka under the Law because they do not come into man's use, nor are eaten by men. Hora (1950) explained the significance of this protection. According to him the misshapen fishes such as eels, globe fish and skates are still not eaten by the orthodox Hindus. In fact, the globe fishes are poisonous and sharks are carnivorous. On that account their eating is prohibited.

Ashoka's Fishery Legislation

After declaring the five fishes mentioned above as inviolable, Ashoka legislated for the conservation of the fishery as a whole. Hultzsch (1925) has given free translation of the relevant passage which runs as follows:

Fish are inviolable and must not be sold on the three Chaturmasis, (and) on the Tishya full moon, during three days, *viz.* The fourteenth, fifteenth and the first tithi and invariably on every last day.

'And during these same days no other classes of animals which are in the elephant part and the preserves of the fishermen must be killed'. Buhler (1894) rendered this passage as:

'At the (full moon of each) of the three seasons and at the full moon of Taisha or Pausha, fish shall neither be killed nor sold during three days, *viz.* the 14th, 15th, and the first of the following fortnight, nor constantly on each last day.'

According to Buhler (1894) 'Chaturmasi is the full moon of each term of season of four months (summer, rains and witner) and it is not possible to decide with certainty which full moons are meant. They may be those of Phalguna, Ashadha and Karttika or those of Chitra, Sravana, and Margasirasha'. The fourth full moon, which is in the passage is that of Taisha or Pausha (Dec. Jan.)

According to Barua (1943) 'Ashoka's expression 'tisu chatummasisu' cannot but mean the three full moon days that occurred in Ashadha, Karttika, and Phalguna, at the end of three four-monthly seasons and were observed in the Middle country as holidays. In connection with Ashoka's Inscriptions, Barua (1941) observed (Part-II, p.97) that 'Provided that the rhythm is maintained, the cadences are right, the sounds are sweet and appropriate in thyming, and the caesuras come spontaneously, it is immaterial whether certain rules of number and gender are obeyed or infringed.

According to Hora (1950) 'tisu chatummasisu' is also a case of the infringement of rules of number and as such it should be taken to mean the third Chaturmasi.

It is curious to note that these prohibitions follow the liens liad down by Kautilya (XIII,5): the king (in a conquered territory) should prohibit the slaughter of animals for half a month during the periods of Chaturmasya (from July to September), for four nights on the full moon days, and for a night to mark the date of his birth, or celebrate the anniversary of his conquest. He should also prohibit the slaughter of females and young ones as well as castration".

Scientific Analysis of Ashoka's Laws

Ashoka's injuctions regarding fish are based on true scientific principles. Hora (1950) examined these laws in greater details as outlined below:

1. No fish should be caught on the 14th and 15th day of the moon and the first day after the full moon during the period of the 3rd Chaturmasi (Sravana, July, Aguust; Bhadra, August-September; Asvina, September-October, and Kartika, October-November) = 12 days.
2. No fish should be caught on the 14th and 15th days of the moon and the Ist day after the full moon of the month of Pausha (December-January) = 3 days.
3. No fish should be caught on the 1st days-Amavasya or the day before new moon and the Ashtami or the eight day during every fortnightly period of the moon = 12 + 24 = 36 days.
4. No tank fish (animals in the preserves of fishermen should be taken during the above-noted days).

The total number of fish-prohibition days enjoined by Ashoka would thus appear to be 51. Barua (1941) has counted 72 at the rate of 3 days in every half luner month, the first, the eighth, and the full or new moon day. Ashoka prohibited the killing and sale of fish for 56 days (24 + 32).

Full Moon Periods in Relation to Fishery Conservation

The first injuction is not to catch fish during the 14th, 15th and full moon days falling during the commencing from the middle of July to the middle of November. It is well known that the peak breeding period of the principal food fishes in India is July, August, and September. However, Ashoka's prohibition period extends up to the middle of November. The logic behind it is that the spent fish all back to their normal habitats in deeper water or to the estuaries and the sea in case of Hilsa. The young also move down to safer habitats after the rains are over. The young and the weakened spent fishes need protection at this juncture. It is significant to note that in Bengtal, Hindus generally do not take Hilsa after the Durga Pooja (Sometime in October) to the Saraswati Pooja (towards the end of

January). It is also significant that prohibition is restricted only to certain specified days and not to the entire season. The fecundity of major carps and Hilsa is well known. A pair of Spawners, under favourable conditions, can produce millions of young. As such, nearly 6 days restriction during each spawning month is sufficient for the conservation of the fisheries. The significance of the second law whereby catching of fish is prohibited on the 14th and 15th day of the moon and the Ist day after the full moon besides the fast days of the month of Pausha (December-January) is not yet clear.

Fast Days in Relation to Fishery Conservation

In regard to prohibition on all fast days it can be argued that the fish trade will not be affected to any extent and fishermen themselves will be able to observe fasts. Moreover, besides the major carps, there are other varieties of fishes which do not breed during the rainy season but at other times of the year. The principal food fishes of the Gangetic estuaries, such as Mullets, Prawns, Bhetki, etc. breed in March and April. The salt-water Bheries take in water containing the eggs and young-ones of these species during the spring tides of the new moon and full moon periods for stocking purposes. The details are contained in the first Fishery Development Bulletin of the Government of Bengal by Hora and Nair (1943).

Prohibition of Tank Fishing in Relation to Fishery Conservation

The law which prohibits tank fishing is the most ingenious of all, for it has nothing to do with the spawning of fishes. It was intended to have a control on the catching of all sorts of fishes found in ponds as well as on the marketing of these fishes. As the tank fishes are the same as are found in rivers, it would have been difficult to control their sale and at the same time prohibit the catching of fish in the rivers.

Evolution of Fishery Laws in India

From the foregoing account, it is evident the Ashoka's pillar edict-V records an advancement for knowledge over what Kautilya had recommended in his Arthasastra (XII, 5) about 25 to 50 years earlier. The scientific basis of Ashoka's laws clearly indicates that they are more perfect and easy to comply with. The grounds for such a contention are that:

1. Ashoka's laws were very simple, applicable throughout his kingdom, and the prohibition periods were evenly spread over the whole of the year thus entailing no hardship either on the consumers or on fishermen. The present day laws are complicated, and through total prohibition during certain season inflict many hardshisps both on the consumers and fishermen.
2. Ashoka's laws were based on the proper understanding of the migratory and spawning movements of the principal food fishes of India.

On the careful consideration of the Ashoka's laws, Hora (1950) was of the opinion that the Indian Union cannot, under present circumstances, think of any better legislative measures for the conservation of its inland fisheries than to enact the laws promulgated by the good king Ashoka in 246 B.C.

Inland Fish and Fisheries in the Jataka Sculptures (200 B.C.)

Indian sculpture is largely an expression of the religious beliefs of the people. Fish are, however, found in reliefs illustrating the Jataka folk-tales, either in connection with the theme of a story or to indicate some points in its artistic treatment.

The remains of a Budhist stupa similar to that of the great stupa at Sanchi, was discovered by General Alexender Connugham in 1873 in a small village Bharhut located in M.P. The railing and the gateway are carved in low and flat relief. Among these carvings, we have a few panels and modallions with representations of fish. According to Hora (1955), the following representation of fishes have been made in six Bharhut reliefs dating back to the second century B.C. and one Sanchi relief of the first century B.C.

Naga Jataka

The scene shows four fishes of which two in full side view are *Catla catla* (Ham) according to their body form. However, the subterminal mouth is characteristic of a Rohu (*Labeo rohita*). The third fish which appears to be either Catla or Rohu, has its head portion in the beak of a crane or a swan, which is in the act of swallowing it. The head portion of the fourth fish, also of the Catla-

Rohu, type is represented as protruding out of a shelter. The topmost specimen may represent a salamander though it does not possess scales. The only Indian species of salamander is found in the eastern Himalayas.

The Jataka text refers to the story as Kakkata (crab) Jataka (No. 267), as the crab is one of the principals of the story.

Uda Jataka

The scene shows the head and tail portions of Rohu (*Labeo rohita*). The fish shown moving about in the river also represent Rohu-Catla type. The fish cut in three pieces is a Rohu. The story is entitled Dabbhapuppha (name of the Jackal) Jataka (No.400).

Mahakapa Jataka

The scene shows three fishes which move in the same direction though one is shown upside down. These fishes represent the Rohu-Catla type of carps.

In the Sanchi relief made a century later, a varities of fishes and other aquatic animals have been shown (Plate 4). They are so crudely represented that it is difficult to identify them.

Kuranga Mrga Jataka

The scene presents a large tree growing in the thicket near a lake. The fishes shown in the lake are carps and represent *Catla* and Rohu.

Makura Medallion

The medallion shows a makara (crocodile) with its mouth wide open and a fish moving freely into it. The fish has been shown into the three positions: the lower one moving freely below the lower jaw of the crocodile, the middle one in the crocodile's jaws and the upper one moving about freely after coming out of the jaws.

Timingila Jataka

The medallian presents a scene in which the sea monster Timingila swallows smaller but similar marine shoaling fishes known as timi. The probable marine shoaling fishes represented in the Timingila Jataka have been identified by Hora (1955) as: Mugil keleartii (Gunther) (*Glupea kanagurta* (Bleeker) and *Therapon qurdrilineatus* (Bloch).

The Jataka sculptures clearly indicates that Rohu and Catla were most popular fresh water fishes in the Bharhut sculptures are sufficiently accurate to permit identification. It is well known that sharks are voracious in nature and Timingila timingila-gila or Jataka might represent the tiger shark (Galeocerdo) and the Whale sharke (*Rhincodon typicus*) respectively.

Fisheries in the Sangam Literature

The kingdoms of south India were distinct from those of North on account of the barrier of thick forests. Three importance kingdoms were recognized *viz.* The Pandyas, the Cheras and the Cholas. The sources of information about the economy and social life of the people of South India is the sangam literature which was complied in the period (300–600 A.D.). The Sangam was an assembly of Tamil poets held under the patronage of Pandya king at Madurai. A few species of fish are referred to in the Sangam literature of south India. There are singularly few references to fish although some inferior kinds like Ayirai, Iravu, Iral, Suyal and the Shark (Suravu) are stated to bear the fish symbol as the Chola and Chera coins bear other animals (vide Dandapany, J.S. in the Hindu, 14th March, 1957).

Fish and Fisheries in the Gupta Period (350–550 A.D.)

The Gupta dynasty was founded in 300 A.D. The poets of this period, Kalidasa deserves to be particularly mentioned. In his master piece. 'Abhigyan Sankuntlam' we find that Dusyanta and Sankuntla were reunited through a lost ring obtained from fish. We get the following passage regarding fish and fisheries in the sixth act of Abhigyan Sankuntlam.

Chapter 4

"FISHERY IN MEDIEVAL AND BRITISH COLONIAL PERIOD"

Fisheries in the Medieval Period

From the 7th to 12th century AD, the fisheries declined. We get only one reference of fish production in Rajpur kingdom of North India.

Fisheries in Bengal

In Bengal, fish production during the reign of Pal dynasty was well known. From a copper inscription of Devapal (810–850 A.D.) in which a moshika was made gift with all conceivable right of

"स्वसीमा तृणयुतिगोचर पर्यनत सतन्न सोद्देश :

साम्र मधुकर : सजलस्थल समत्स्य : सतृण . . . "

Mention of three distinct crops are made here *i.e.* mango, mahua and fish.

In the next 3 to 4 centuries, there was a revival of interest in fisheries as evidenced by the epigraphical records of the Pandya and Chola kings regarding irrigation project during 1000 A.D.–1600

Venkayya (1960) in his article on irrigation in southern India has referred to a large number of irrigation tanks mentioned in Pallava, Chola and Vijaynagar inscriptions. From the following 7 Tamil inscriptions ranging from 1378 A.D. to 16th century A.D. it would appear that the income from the lease of fishery tank was generally sufficient for maintenance and improvement of irrigation tanks.

Inscription No. 164 of 1943

Chivvada, Truttani Division, Chittoor District Stone in the Village (AD 1378). Records, in Pramadi year during the reign of the Vijaynagar king Hairhararaya, that Mahammandalesavara Macha Reddi Ganpati-Red-diyar constructed a sluice for the tank at Siyapadi alias Uttamasolanallur and that he provided for its dredging by authorizing the fish lease money derived from it to be utilized for the purpose.

Inscription No. 167 of 1944

Pandravedu, Chandragiri taluk, Chittoor district. Stone set up on the tank bund near the village (A.D.1409). Records, in Saka 1332, in the reign of the Vijayanagar king Devaraya, an order of Mallamanna Udaiyar by which Madiarasar made arrangements to spend the amount realized by sale of fish in depending the tank at Pakanravedu.

Inscription No. 168 of 1942

Indiravanam, Polur taluk, North Arcot district. Slab setup on the tank bund (A.D.1445). Record in Saka 1368, Krodhana in the reign of the Vijaynagara king Devaraya Maharaja, the provision made by Tammu Nayakar for dredging the tank at Indiravanam with the proceeds of the sale of fish acquired from the tank.

Inscription No. 145 of 1924

Kodungular, Wandiwash taluk, North Arcot distict. On a slab lying in the tank (A.D. 1521). Records, the gift of the income from the lease of fishery in the tank at Kodungalur in Marudadunadu for deepening the tank by Dalavay Sevvappa Nayaka, for the merit of Tirumalai Navaka, the agent of the Vijayanagara king Krishnaya Deva Maharaja in the year Vikrama, Tai, 5 (A.D.1520).

Inscription No. 262 of 1928

Nanguneri, Nanguneri taluk, Tinnevelly district, North wallk of the outermost prakara of the Vanamalai-Perumal temple (A.D.1591). Records, in Saka 1513, Kollam 766, Khara, the grant by the temple authorities, and the assemblies of Sivaramangai etc. of the right of fishing (pasi) in the Solapandyaperi to certain residents in return for clearing the silt of the tank every year.

Inscription No. 16 of 1945

Puliyar, Teruvallur taluk, Chingleput district. Slab in the Perumal temple (16th century A.D.). States that a certain Madaliar, the agent of Tirumalainayakkar assigned the fish income from the tank at Puliyur for the expenses of dredging and strengthening its bund.

Inscription No. 97 of 1943

Sitteri, Polur tal, North Arcot district. Stone lying in the dry lake-bed (16th century A.D.). Fragment. Seems to state that the income from the fish lease should be utilized for the upkeep of the lake.

Inscription No. 266 of 1942

Madura district First group of the Sundaresvatra temple. Highly damaged, date etc. lost. Seems to record gift of land to a bhatta of the Asvalayana sutra and refers to the spending of the fish lease amount on the local tank.

From this account it is evident that in the first five inscriptions (A.D. 1378 to A.D. 1591) the fish lease money was sufficient for desilting the tanks but in the later three inscriptions it is stated to have been used for dredging the tank and strengthening its embankments.

The mention of fisherman in the middle of the 10th century A.D. in connection with the tank at Nangavarm is of special interest. Between the 11th and 14th century, the practice of stocking and rearing of fish species of economic importance in irrigation tanks must have sufficiently advanced so as to produce enough revenue from the fishery tank for its proper maintenance. The fisherman seem to have supplied wood, presumably for repairs to the boat, and for this service received 2 kanals of paddy annually. The carpenter and blacksmith, who were part-time workers for the tanks, got 2½ kanals of paddy annually as against the full time workmen who got one padakku of paddy per person per day. It appears reasonable to presume that the normal occupation of the fishermen was to fish in the tank and that in addition they supplied wood for repairs to the boat.

Art of Angling

The text of 'Matsyavinoda' in 'Manasollasa' written by king Somesvara in 1127 A.D. is the first known treatise on the art of angling in India. King Somesvara was the son of king Vikramanka or Vikramaditya VI, the greatest of the later Chalukyas, who patronized learning and literatures in India. Chapter XIV of Manasollasa shows details unrivalled even in modern works on the art of angling. Somesvara had the tile of Sarvajana-chakravartin, the 'omniscent emperor' or the 'emperor among all knowing'. This would show that he had acquired a knowledge of many arts and sciences.

Marine and Estuarine Forms

The marine species are divided into scaleless and scaly and further subdivided according to size. Six scaleless marine fishes are mentioned and all of them are stated to be of large size. These are as follows:

1. Sora = Sharks
2. Sringasora = Saw-fish, a kind of shark
3. Chambilocha = Hammer-headed Shark
4. Bala = Cat fishes of the genus Arius
5. Kantakara = Cat fishes of the genus Plotossus
6. Sakuchaka = Skates and Rays

Three species have been referred to amount the scaly marine fishes, one of which (Pandimana = *Lates clacarifer* or Bhetki) grows to a large size, while the other two (Pallaka = The red Rock Perch, *Lutianus rouses* and Tomara = The Gar Fish, *Belone annulata*) are stated to be medium size.

Kaurattha and Kovakiya have been listed among the estuarine form. Abut the former, King Somesvara says:

"Along the mountain bordering the Mahanadi, Kaurattha comes six to seven yojanas from the sea to river. They reside in large deep pools. This itself, and not the sea, is the place where to catch them. They live in the river where there is rocky bottom and rocky sides and no mud.

The identification of this fish is made difficult as Mahanadi is outside Somesvara's Kingdom and that there are no mountain bordering the Mahanadi within 60 to 70 miles from its mouth. Hilsa, however, seems to satisfy two main points in the description of Kaurattha, namely (*i*) it swarms up the river Mahanadi in large numbers for a considerable distance up-river and (*ii*) its main fishery lies in the river and not in the sea.

Kovakiya is stated to spend part of its life in the sea and part in rivers. 'They live in deep waters studded with rocks'. Certain species of *Polynemus* and *Sciaena* enter estuaries for breeding purposes, so it is likely that they are referred to in Kovakiya, the etymology of which is unknown. It is interesting to note that Macdonald (1948) has also listed *Lates calcarifer, Lutianus roseus, Polynemus tetradactylus, P. indicus*, rays, sharks etc. among the sea and estuarine sporting fishes. Col. Burton gives an account of Sea fishing along the West Coast of India. On page 219 he listed practically all the species enumerated by Somesvara. Among others, he mentions Ground Shark (*Charcharias gangeticus*, Hammer-heads shark (*Eygaena blockii*), skate (*Rhynchobatus djeddensis*), Marine Catfish (*Arius spp.*), Sting ray (*Trygon sephen*), Garpike (Belone sp.) and Perches (*Lates, Kunianus*), etc.

It would thus appear that King Somesvara was influenced by the abundance of the marine species and by their sporting qualities. He was aware of the fact that 'Fishes are of many kinds: they defy enumeration', but he dealt with some of the sporting fishes only.

Freshwater Riverine Forms

Somesvara mentions 21 species, of which 6 are scaleless and 15 with scales. Scaleless species are further divided into large (4) and medium (2) size. Of the 15 species with scales, 5 are of large size, 6 of medium of size and 4 of small size. The various species are listed below with their probable scientific names.

Scaleless Riverine Fishes of Large Size

1. Kovasaka = *Mystus aor* (Ham.) of *M. seenghala* (Sykes)
2. Khirida = *Pangasius pangasius* (Ham.)

3. Pathiana = *Wallago attu* (Bloch).
4. Simhatundaka = *Bagarius bagrius* (Ham.)

Scaleless Reverine Fishes of Medium Size

1. Patalapichchhaka = *Clupisoma garua* (Ham)
2. Dantapatala = *Eutropiichthys vacha* (Ham.)

Large-Sized Riverine Fishes with Scales

1. Mahasila = Mahasirsha or Mahasiras = *Tor mussulah* (Saykes) and *Tor khudree* (Sykes). The famous sporting fish mahseer.
2. Kahlava = *Barbus carnaticus* (Jerdon)
3. Nadaka = *Barbus curmuca* (Ham.)
4. Badisa = *Notopterus chitala* (Ham.)
5. Vatagi = Not identified

Medium Sized Riverine Fishes with Scales

1. Rohita = *Labeo fimbriata* (Cuv. & Val.)
2. Svarnamina = *Barbus sarana* (Ham.)
3. Khandalipa = *Mastacembelus armatus* (Lacel.)
4. Marila = *Ophicephalus striatus* (Bloch.)
5. Vanchi = Not identified

Small-Sized Riverine Fishes with Scales

1. Gagdhara = *Xenextodon cancila* (Ham.)
2. Gojjala = *Ophicephalus punctatus* (Bloch.)
3. Vidruva = Carp-Minnows (*Puntius, Danio, Esomus*, etc.)
4. Kantharaya = *Barilius benidelisis* (Ham.)

The following species enumerated by Somesvara have not been identified:

1. Vatagi = Scaly, riverine, large size = Vataka or disc-like globular fish
2. Vanchi = Scaly, riverine, medium size; meat eater, lives in pools

3. Koraka = Lives in deep sandy pools
4. Thogyra = Lives in pools

Fish Tackle

The tackle for angling is dealt with by King Sameswara under the heading-line, hook and rod. In each heading, a fairly detailed instructions were given by him.

The Line

Four types of material, namely, fibres of marva or the Bow-string hemp (Sansevieria Zelanica Wild and *S. roxburghii*); fibres of Kanduka or Kandula of the Maratti (*Sterculia urens* Roxb.); fibres of Arka (*Calatropis gigantean*) and fine threads of cotton are suggested to be used for the line. The line made out of murva fibres is the best. The next in quality are those made from the fibres of Kanduka or Arka. The line made of the cotton thread are of the poorest quality soft and silky, is evident form the fact that it is much valued in Europe for making ropes used in deep-sea dredging.

He divides the line into three categories–thick, medium and fine. He wants them to be three-ply, long and soft to touch. They should be twisted by turning of fingers. The line should in no case be longer than 300 yards and shorter than 3 yards. The extreme thickness of the line may be 1/6 or 1/10 inch and the thinness like the hair-from a horse's tail. It is indeed remarkable that Macdonald (1948) also gives 30 yards of line as the maximum length. It seems probable that the evolution of the modern line has been due to the use of the reel, which was not known in Somesvara's time.

The Rod

The rod made of bamboo, sprung from the earth is considered to be the best. The next in quality is a spike-shaped branch of a bamboo with close joints and solid from inside. King Somesvara also recommends a branch of the Mada tree (Mangrove tree or Avicennia). The rod should be tapering like the tail of a lizard. In the case of a thick bamboo rod, its circumference at the thickest point should be about 3 inches (six fingers) and in case of a small rod, half the thickness. The rod should neither be very long nor very short; neither very supple nor very stiff. In the case of a Mada tree rod, its greatest circumference could be about 5 inches (ten fingers). It should taper gradually and the joints should not be close together. Macdonald (1948), writing about rods, says:

"The old idea of long sixteen feet rods, heavy spoons, and baits, have given place to ten and twelve feet rods, now almost universally used for the heaviest fish".

The Hook

The hook should be made of iron, which is bent wide in the middle with the base thick and point thin. The part where the line is tied may be flattened or rounded. The curve of the hook should either correspond to a goad, or the hoog of a horse, or like a mango or crab-shaped. It must have an additional spike near the sharp end. Macdonald (1948) also mentions that in a hook 'The important thing is a short shank with a sharp point, and a thin short barb'.

Fish Baits

Baits are used to attract fishes to particular spots where angling operations are to be done. As different fishes have different feeding habits, naturally the ground baits prescribed by Somesvara are also different. Different types of baits are listed below according to the varieties to fish to be fed.

1. For Kahalava (*Barbus carnaticus*) and other similar fishes, the ingredients like cakes and balls of Sesamum seeds, powdered or parched rice, flour of parched gram, cooked rice should be used. The ingredients are mixed and balls of 2 to 2½ inch size are to be made. These ingredients can also be powdered with oily substances and then thrown in the stirred water.
2. For Rohita (*Labeo fimbriata*) and other similar fishes, sesamum-cake or seeds are to be mixed with cooked rice and balls of the size of a plum are to be made out of it.
3. For Badisa (*Notopterus chitala*). The power of parched sunflower and flour of barley are mixed with cooked rice and make balls of the size of a mango.
4. For Vatagi, Crushed leaves of the wood-apple tree mixed with parched barley and made into balls as big as amla fruit are suitable for feeding this fish.
5. For Kauvaka (*Polynemus tetradactylus*)–Bits of lungs and muscles.

6. For Pathiana (*Wallago attu*)–Bits of foul-smelling meat.
7. For Simhatunda (*Bagarius bagarius*)–Dhichakas (meaning unknown).
8. For Marali (*Ophicephalus striatus*)–Crab flesh.
9. For Turtles–Roasted meat of mice.
10. For small fish–Earthworms.

Regarding ground baiting, Macdonald (1948) writes as follows

'There are many kinds of ground bait; oil cake either mustard or castor, fried or fermented is good. Mix it with mud or gram or atta or rice, and throw it in.' It can be concluded that Somesvara shows a greater insight into the feeding habits of the fishes by dividing them into groups and then prescribing a suitable ground bait for each group or kind. Further, Somesvara also prescribes the limit in size for the morsel of each variety of fish. He wants the ground bait to be prepared into balls. While Macdonald (1948) gives proportions of the various ingredients to be mixed, Somesvara gives no such indications in his account.

Somesvara's Advice on Fishing

Ground Bait

He considers regular ground baiting as a necessary condition preceding actual fishing.

Tackle

The examination of the tackle to be used for fishing is Somesvara's next instruction. For instance, one should not simply tie a thick cord to a thick rod or a thin cord to a thin rod. One should determine the length and thickness by taking into consideration the depth of the water and the size of fish. One should attach to one end of the cord a fishing hook, the size of which should be consistent with the length of the cord.

A Quill Float

A peacock feather should be tied in the middle of the cord. At the base of the rod another long cord should be tied. For the purpose of attracting the fish, the expert should make the rod and the line appear like a tail tapering to a point.

Bait

Flour or meat, whatever is liked by a particular fish desired for spot, should then be put on the hook as a bait.

Casting the Line

Having attached the bait to the hook, one should throw the baited hook at the place where usually food (ground bait) is thrown to the fish.

Fish Bite

Having cast the line, the angler should keep looking at the feather attentively. As soon as the fish touches the bait and begins to eat it, there is movement of the feather. This movement is different in the case of different fishes-meat-eating and flour eating kinds.

Striking a Fish

As soon as the movement of the feather is noticed, the angler should strike with a sharp pull.

Playing a Fish

When a fish gets entangled on the fishing hook, it becomes very vigorous. As soon as it gets exhausted, one should haul it up. If it shows signs of vigour, one should release the cord. Gently and slowly one should haul it up, taking care that the cord does not give way. If the rod has to be thrown after the fish, one should withdraw it by the tail string.

Thus it seems probable that Somesvara's Matsyavinoda probably contains the accumulative knowledge about fishes of the later half of the 10th century, the whole of the 11th century and nearly three decades of the 12th century. In any case, the knowledge recorded is remarkably accurate and instructive and could be used in pond cultural practices to which considerable attention is now being paid in India. For the fattening of fishes, Somesvara's ground bait recipes could be tried with advantage.

Fisheries in the British Colonial Period

For an organized study of fish and fisheries in India, we owe much to British residents of India, some of whom where keen naturalists and sportsmen. We also owe particularly to the learned societies, like the Asiatic Society of Bengal in Calcutta and the Bombay

Natural History Society in Bombay which they helped to establish as the representatives of British ruling class.

In the Historia Piscium published in Oxford in 1686 (edited by John Ray after Francis Willoghoby's) 420 species of Indian fish were recorded. Petrus Artedi (1705–1735), who has been justly called the father of Ichthylogy was born in Sweden and was a fellow student of Linnaeus at Upsala. He went to Holland to examine a large and valuable collection from the East Indies belonging to a rich Dutch merchant in Amsterdam named Albert Seba. At the age of twenty-nine he was drowned there as the result of an accident in one of the Dutch canals. His manuscript work was fortunately rescued and was edited and published by Linnaeus in five volumes. Artedi attached a descriptive phrase, consisting often of a number of words, to the name of the genus to make up the name of the species. This system of nomenclature is called polynomial.

L.T. Gronow, a German, who resided in Holland, closely followed the arrangement proposed by Artedi and increased the number of genera were described from types obtained from the Coromandel coast. His well known genus *Ophicephalus* is one of these. His lchthyologia was published in Berlin between 1782 to 1785. After the completion of this work, Bloch occupied himself with systematic classification of fishes, in which he arranged and described a large number of genera and species of fishes. After his death, the work was edited and published in 1801 by J.G. Schneider under the title "M.E. Bnlochii System Icthylogiae". The work contains 1519 species of fishes. Though Bloch had formed many Indian genera in his "History of Fishes", but a much more considerable number was formed by Lacepede in his continuation of Buffon's Natural History. The great work of Comte de Lacepede, Histoire Naturellel des poisons, was published originally in five volumes in Paris between 1798 to 1803.

In the later part of the eighteenth century there flourished two very enthusiastic workers in Indian Ichthylogy. One of them was Dr. Patrick Russell of the Madras Medical Service and the other Dr. Francis Hamilton (formerly) Buchanan of the Bengal establishment.

Dr. Patrick Russel in course of several years of his stay at Vizagapatnam was attracted to Ichthyological research by the sight of fishermen dragging their large seine nets and angling from boats and rafts beyond the surf. He went back home with two hundred figures of fish faithfully delineated by an Indian artist under his supervision. These figures were published with detailed descriptions in 1803 under the patronage of the East Indian Company at the suggestion of Sir Joseph Banks under the heading "descriptions and Figures of Two hundred Fishes collected at Vizagapatam on the Coast of Coromandel".

Dr. Francis Buchanan (1762–1829), came from Edinburg and entered the service of the East India Company as an assistant surgeon on the Bengal establishment in 1794. He was a keen student of nature before his arrival in India. His enthusiasm continued throughout his official life in this country. He made several voyages in the Western Sunderbans and utilized these opportunities to become acquainted with estuarine fishes. In 1800 he was commissioned to report upon the state of Mysore and Malabar, lately acquired from Tipu Sultan. In the course of these travels he discovered three new species of fish which were included in his reports published in three volumes between 1805-07 under the title "A journey from Madras through the countries of Mysore, Canara and Malabar, etc.". These descriptions are the first published contribution by Buchanan to Ichthyology. His copies notes on fishes collected while he was stationed at Puttahaut and at Bauripur in 24 Paragana district of W.B. were incorporated in his "Account of the Gangetic Fishes" to be referred later. In 1802 he was sent with Captain Knox to Nepal where he collected a number of fishes.

In 1803, he was appointed surgeon to the governor- general Lord Wellesley. He continued his study of Gangetic fish when he had charge of the manager at Barrackpur, (15 miles north of Calcutta) during 1804–1805. In 1806, Court of Directors authorized him to make a comprehensive statistical survey of the territories comprising the presidency of Bengal as well as a portion of Assam and some other adjacent districts. Buchanan's time was wholly occupied in this work for seven years from 1807 till 1814. He covered the districts of Rangpur, Dinajpur, Gopalpara, Purnea, Bhagalpur, Monghyr, Gaya, Patna, Shahabad, Gorakhpur and even Allahabad and Agra also.

Buchanan took extensive notes on fish and fisheries of the districts he surveyed and incorporated them in the report which he submitted to the Court of Directors of the East India Company. Verbatum extracts from this manuscript report were published by

Montgomery Martin in 1835 in three volumes under the tile "Eastern India" with Martin's name on the title page, a work popularly known as Martin's India. The full report on fish and fisheries was eventually published by Sir William Hunter in 1877 in the twentieth volume of his "Statistical Account of Bengal" with footnotes by Dr. Francis Day on the names under which the fish are referred to in the "Fishes of the Ganges". The drawings illustrating Bnuchanan's work, however, have not been published as yet. One set consisting of one hundred and forty-nine original coloured figures and forty-five copies of originals, all made by Indian artists under Buchanan's supervision, are in the possession of the Asiatic Society of Bengal.

In 1815, Buchanan returned to England (home) for the last time and both his parents having died, he fell heir to the extensive property of his mother (Amiss Hamilton) and hence he assumed Hamilton as the surname. For a number of years he was known to English and Indian writers as Francis Buchanan Hamilton.

On his return an "Account of the Fishes found in the River Ganges and its Branches" appeared in 1822.

In the library of Asiatic Society Bengal several interesting works have been rediscovered. Hora (1926a) has found four volumes of Buchanan's Zoological Drawings in the library of the Asiatic Society of Bengal, two of these contain fish drawings.

Volume 1 contains 20 plates of fish drawings

The following table gives the distribution of drawings on these plates:

Serial Number of Plate in the volume	*Serial Number of Fish Plates*	*Number of Coloured Drawings on Each Plate*	*Number of Complete outline Drawings on Each Plate*	*Number of Species Represented on Each Plate*	*Number of Drawings of Part of Fish on Each Plate*
96	1	1	1	1	
97	2	1	2	2	
98	3	1	1	1	
99	4	3	2	3	
100	5	3	0	2	
101	6	4	1	3	1
102	7	3	3	3	1

Serial Number of Plate in the volume	Serial Number of Fish Plates	Number of Coloured Drawings on Each Plate	Number of Complete outline Drawings on Each Plate	Number of Species Represented on Each Plate	Number of Drawings of Part of Fish on Each Plate
103	8	2	2	2	1
104	9	3	2	3	
105	10	4	4	4	
106	11	4	4	4	
107	12	2	2	2	
108	13	3	3	3	
109	14	2	2	2	
110	15	2	1	2	1
111	16	2	2	2	
112	17	3	1	2	
113	18	3	3	3	
114	19	5	2	4	1
115	20	3	3	3	
		54	41	51	5

Thus there are 100 separate drawings in all, of which 54 are coloured illustrations, 41 outline drawings (dorsal view of fish) and 5 sketches representing some important structures. Of the 51 species represented on these plates *Mugil bongon* is figured in three places *viz.* on plates 98, 99 and 114. Volume IV contains only fish drawings and includes all those fishes listed by Day. Hora (1918) cleared up the discrepancy observed in the counting of these drawings by McClelland and Day.

Day examined and listed in the IVth volume 146 drawings of which 64, 70 and 71 refer to the same species *Mugil bongon*. Thus the total number of species examined by day is	144
Day missed out one drawing 125a in numbering	...1
Drawings which occur in volume 1 but those original are missing in volume IV	5
Total number of species illustrated in Buchanan's Collection	150

After Buchaan's departure from India his Zoological drawings were placed under the charge of the Superintendent of the Royal Botanical Garden Dr. Wallich in November, 1924. Permission was obtained by Dr. Baini Prasad from the Council of the Asiatic Society of Bengal for copies to be made of all the fish drawings in volume IV for the Library of the Zoological Survey of India. He also undertook to have the original drawings properly bound up and preserved. The number inserted on the drawings in pencil by Day have been printed and pasted in their places and also the published names as deciphered by Hora (1926). The volume, however, presents an absolutely different appearance, from what it did before the binding.

No. of Drawings	*Unpublished Names as Given by Day*	*Unpublished Names as Made out by Hora (1926)*
4	Silurus chake	Silurus chaka
6	Silurus panda	Silurus pabho
14	Pimelodus muri vacha	Pimelodus muri bacha
21	Pimelodus cavasi	Pimelodus cabasi
23	Pimelodus viridescens	Pimelodus Viridescene
24	Pimelodus ? mangra	Pimelodus nangra
27	Ophisurus rostralus	Ophisuris rostrata
35	Squallus characias ? karnia	Squallus characias
41	Acheris jibha	Acheiris jibha
42	Acheiris kukar jibha	Acheiris kukars jibha
77	Makalkar	Makalpar
79	Sygnathus deokhuta	Signathus deokhata
80	Syngnathus kharke	Singhathus kharke
82	Polynemus	Polynemus tetradactylus
83	Polynemus raye	Polynemus raye
85		Cottus grunniens
98	Cyprinus pungsi	Cyprinus paungsi
106	Cyprinus sada balitora	Cyprinus sada balitora Desari
119	Cyprinus kursi	Cyprinus kursi Nathpur
126	Cyprinus goha	Cyprinus goha vaghra

The Manuscript Drawings of Fish in the Library of the A.S.B. (Mackenzie Collection 1784–1814)

Lt. Col. Colin. Mackenzie made a valuable collection of oriental manuscripts and articles from southern India at the time when the two pioneer Indian Ichthyologists, Patrick Russell and Francis Buchanan (afterwards Hamilton), were busy collecting materials for compiling the Fishes of 'Vizagapatam' and for 'An Account of the Fishes found in the River Ganges and its branches respectively'.

In his descriptive catalogue of the Mackenzie collection, H.H. Wilson gives a list of drawings and "portfolios". In the list of drawings are mentioned 65 originals and 58 duplicates of natural history objects. Among the portfolios are mentioned 65 original drawings of natural history objects. At present this volume contains 90 drawings, on 55 coloured plates, which are distributed as follows:

Insects 02
Spider 01
Fishes 26
Reptiles 08
Birds 50
Mammals 03

The list of fish drawings from the Mackenzie's collection is gien below:

List of Fish Drawings in the Mackenzie Collection in the Library of the Asiatic Society of Bengal

No. of Drawing in the Collection	*Vernacular Name*	*Probable Scientific Name*	*Locality*	*Date of Collection*	*Length of Specimen*
1 & 2	Gooly–Champi (Can) or Blind fish	*Bagarius yarrellii* (Sykes)	Hooggree river near Ballary	24.3.1801	3 ft
3	Illegible	*Barbus* sp.	Hurryhur	12.9.1800	11 inches
4	Callarnah (Hind)	*Aoria seenghala* (Sykes)	Hoggree River near Ballary	20.3.1801	2 ft.

No. of Drawing in the Collection	*Vernacular Name*	*Probable Scientific Name*	*Locality*	*Date of Collection*	*Length of Specimen*
6	—	*Callichrous malabaricus* (Cuv. & Val.)	–do–	–do–	1 ft. 3 inches
7	Seepry	*Barbus* sp.	–do–	–do–	2.5 inches
8	—	*Wallago attu* (Bloch & Schn.)	–do–	–do–	6.8 inches
9	Seepry	*Ambassis* (H.B.)	–do–	–do–	2 inches
10	—	*Barbus* sp.	–do–	–do–	2.75 inches
11	Forghee (eatable)	*Aoria aor* (H.B.)	–do–	–do–	9 inches
12	Guirlah	*Aoria cavasius* (H.B.)	–do–	–do–	6.4 inches
13	Corcha	*Gagata itchkeea* (Sykes)	Toombudra River	–do–	—
14	—	*Chela boopis* Day	–do–	–do–	—
15	—	*Labodussumieri* (Cuv. & Val.)	–do–	–do–	—
16	Belchau	*Aspidoparia* sp.	–do–	–do–	3.5 inches
17	Gungatally	*Rohlee cotio* (H.B.)	–do–	–do–	3.5 inches
18	Chidevay	*Callichrous* spp.	–do–	–do–	5 inches
19	Chittra–Mine (Can.)	*Labeo nigrescens* Day	Hoggree River near Ballary	26.3.1801	1 ft. 5 inches
20	—	*Mastacembelus Armatus* (Lacep.)	Cubbanee River	11.9.1804	1 ft. 1 inch
21	—	*Ophiocephalus marulius* (H.B.)	–do–	—	7.25 inches
22	Frog Fish	*Tetraodon nigropunctatus* (Bl. & Schn.)	Sea coast near	January, 1794	—
	—	*Kanduka* sp.	St. Thome River or back water	2.12.1814	—

Almost contemporaneous with these workers in India, a mastermind in France revolutionized our conception of Ichthylogy, India. It was Georges Leopold Chretein Frederic Dagobert Cuvier (1769–1832), with whom we have the beginning of a new era in this branch of Zoology. The great naturalist, Cuvier was familiar in a general way with the fisheries of India through the correspondence of friends residing in this country.

Cuvier did not occupy himself with the study of fishes merely because this class formed part of the "Renge animal arrange appresson Organization (1817)" but devoted himself to it with particular predilection. Indefatigable in examining all external and internal characters of fishes from all parts of the world brought together through his ceaseless activities and masterful influence, he ascertained the natural affinities of the infinite variety of species and accurately defined the divisions, orders, families and genera of the class. Soon after the year 1820, Cuvier, assisted by one of his pupils, A Valenciennes commenced his great work on fishes, Histoire Naturelles des Poissons, of which the first volume appeared in 1828. After Cuvier's death in 1832 the work was left entirely to Valenciennes who left it unfinished with the twenty-second volume (1848). Incomplete as the work is, it is indispensable to students of Indian Ichthylogy though Cuvier had to depend too much on the report of M. Leschenault, a superficial observer, and his collections from Pondichery, and often, disregarded or passed over the results of Hamilton Buchanan's masterful studies. It has been already noticed that the great work of Cuvier and Valenciennes was left incomplete; but several authors subsequently supplied detailed accounts of the order omitted in that work. Johannes Muller (1808–1858) of Berlin with Dr. J. Henle published an account of the Plagiostomes and J.J. Kaup of the Muraenida and Lophobrancbi. A Dumeril commenced on Histoire Naturelle des Poissons ou Ichthyologi Generale of which only two volumes (Plagiostomes and Lophobranchi) were published, when the publication was suspended owing to the death of Dumeril. All these publications are of paramount importance of the students of Indian fishes, nor can they afford to neglect the results of various voyages of discovery mostly stimulated by the work of Cuvier and his disciples.

Major General Hardwicke (1798–1834) was making an extensive collection of natural history specimens with the aid of which John Edward Grey published his illustrations of Indian

Zoology between 1830 and 1832. Fish drawings from the Indus and Kabul rivers were made about the year 1836 by P.B. Lord who was a naturalist in the mission carried out by Sir. Alexander Burnes (1805–1841) to Kabul. There are in all 32 species of fish figures from the Indus. These are as follows:

List of the Fish Drawings in the Burnes Collection in the Library of the Asiatic Society of Bengal

No. of Fish Plate	*Local Name*	*Scientific Name as Entered on the Plate*	*Probable Scientific Name*	*Locality*
1	Jirko	Probably *S. anastomus* C.V.	*Wallago attu* (Bl. & Schn.)	Caught at Hydrabad
2	Pulvah (Jelee in pencil)	Bargrus allied to *B. bacha*	*Pseudeutropius murius* (H.B.)	Attock, 5th August, 1837
3	Doongoono	Probably *Schilbe garna*	*Pseudeutropius garua* (H.B.)	Caught by hook in the Arrut at Sehwun
4	Singaro	—	*Macrones aor*	Sehwum
5	Singalah or kaga	*Pimelodus Rita Buch*	*Rita rita* (H.B.)	Hyderabad
6	—	—	Oreinus, probably *plagiostomus* (Heckel)	Fish of Cabool
7	Bottle (Butul in pencil)	*Racoma labiatus* McClell	*Schizothorax Labiatus* (McClell)	River of Cabool and Attock
8	Singari	—	*Schizothorax* sp., probably nobilis (McClell)	Attock
9	Dahee (Kulbis Lahore)	*Cyprinus calbaus* Buch	*Labeo calbasu* (H.B.)	Caught in the Kinjor Lake and at Lahore 3rd July, 1838
10	Rahoo	—	*Labeo dyocheilus* (McClell)	Peshawar
11	The Moraku	—	*Cirrhina mrigala* (H.B.)	Caught in the Delta of the Indus
12	Moree (Mriga, Lahorre)	*Cyprinus mrigala* Buch.	*Cirrhina mrigala* (H.B.)	Hyderabad and at Lahore, 3rd July, 1838
13	(a) Ruhu	—	*Barbus* sp. (H.B.)	Husn Adal Ajson Abdad
	(b) Butur	—	*Barbus* sp.	–do–

No.of Fish Plate	Local Name	Scientific Name as Entered on the Plate	Probable Scientific Name	Locality
14	Londah	—	*Barbus* sp.	Peshawar
15	(a) Goge	*Mastacembelus Rynchobdella*	*Mastacembelus armatus* (lacep)	Killed at Sehwum
	(b) Chilloree	*Silundia*	*Silonida silondia* (H.B.)	Killed at Sehwum
	(c) Panpree	*Leuciscus* (Klein)	*Cirrhina* sp.	–do–
16	(a) Tailee	—	*Catla catla* (H.B.)	Caught in the delta of the Indus
	(b) Poplee (Pooteaea, Darra)	—	*Barbus chrysopterus* (McClell)	–do–
	(c) Giam	*Bagrus Cavasius*	*Mystus cavasius* (H.B.)	–do–
	(d) Moondeea (Daffo.)	—	*Ophiocephalus* probably *punctatus* (Bloch)	–do–
17	(a) ______	______	*Lepidocepha-lichthus guntea* (H.B.)	Khyrabad, Attock
	(b) ______	______	*Barilius*	–do–
	(c) ______	______	*Danio rerio* (H.B.)	–do–
	(d) Kaunul	______	*Barbus* sp.	–do–
	(e) Daku	______	*Ophiocephalus* sp.	–do–
	(f) ______	______	*Danio* sp. Probably *devario* (H.B.) *Barbus* sp.	–do–
	(g) ______	______	*Barbus* sp.	–do–
	(h) ______	______	*Ophiocephalus*	Burru, Attock
	(i) ______	______	Sp.	–do–
18.	Pulla	______	*Hilsa ilisha* (H.B.)	–do–
19.	Gundun	______	*Notopterus chitala* (H.B.)	Hyderabad

W.A. Sykes (1824–1831)

W.A. Sykes, who entered the Bombay Army in 1804 was engaged in statistical enquiry from 1824 to 1831. He made a report on the fishes of South Western India giving description of forty-six species illustratd by twenty eight drawings. It was published in 1841 in the transactions of Zoological Society of London.

Dr. John McClelland (1839–1841)

Dr. McClelleand who was in the Bengal Medical Service continued the work initiated by Buchanan. In 1839, he published a memoir on Indian Cyprinidae in the second part of the 19th Vol. Of Asiatic Researchers (pp.217–465). He also contributed numerous papers of Indian fishes in the pages of the Calcutta Journal of Natural Hisotry which he published and edited for six years.

Dr. Cantor (1839–1860)

Dr. Cantor of Bengal Medical Service wrote a paper entitled 'Notes respecting some Indian fishes' in the Journal of Royal Asiatic Society (1839). He made these observations while discharging the duties of surgeon to the Honourable Company's Marine Survey. Subsequently in 1849–1850 his Catalogue of "Malayan fishes": was published in the Journal of Asiatic Society of Bengal. It contains accurate description of 292 species. His collections became the property of the Honourable East India Company and were transferred to the British Museum in 1860.

Dr. P. Bleeker (1819–1878)

He was a surgeon in the service of the Dutch East Indian Government. Between the years 1840 and 1870. He made a large number of collections of the fishes of the Indo-Australian Archipelago and described them in numerous papers published chiefly in the Journals of the Batavian Society. In 1853 he published a paper entitled 'Ichthyologische fauna van Bengalen', containing almost all the fishes previously described from India with detailed descriptions of 11 speceis of carp from Ceylon. Soon after his return to Europe in 1860, he commenced the publication entitled, 'Atlas Ichthyologigue des Indes Orientales Nerlandaises' with coloured plates. Its publicatios was interrupted after the ninth volume by the author's death in 1878.

Dr. Albert C.L.G. Gunther (1859–1870)

His monumental work, 'Catalogue of the Fishes of the British Museum' was published from 1859 to 1870 as a result of his laborious devoition and patient industry for a number of years. In this work, there are 6,843 speceis of fish described and 1,682 doubtful species reported.

Mr. Blythe (1841–1862)

Mr. Blythe, the distinguish curator of the museum of Asiatic Soceity of Bengal contributed several interesting and important articles to the journal and proceeding of the society from 1858 to 1860 dealing with fishes colleted from Indian and Burma and the Andaman Islands. He also described some of the fishes collected by Dr. Jerdon.

Mr. Grote and Col. Strachey (1876)

In 1867, at the instance of Mr. Grote, a senior member of the Board of Revenue and Col. Stratchey, Inspector General of Irrigation Works of Bengal, made enquiries on fisheries. A very useful questionnaire regarding fish and fisheries of Bengal was issued. It is so comprehensive that it can be regarded as the basis for the future programme of fisheries investigations in the province. Mr. H.S. Thomas of Madras Civil Service has left valuable informations in his books 'Pisci Culture in South Canara' (1870), 'Rod in India' (1873), 'Tank Angling in India' (1887) and 'Tank fishing in India' (1887).

Dr. Francis Day (1826–1889)

Francies Day joined medical service in 1852 and was stationed at Cochin from 1859 to 1862. His work in spare hours resulted in a folio volume on the fishes of malabar in 1865. In 1867 Sri Arthur Cotton drew the attention of the secretary of state to the great harm being done to the coastal fisheries by weirs constructed in all the principal rivers of the east coast for irrigation. The Madras Government appointed Francies Day in 1869 to investigate the matter. His deputation was afterwards extended to Orissa, Bengal and subsequently to British Burma and the Andaman Islands. He reported that the knowledge of the Ichthoyology of this part of the world was exceedingly imperfect. On this part, Dr. Day was appointed Inspector General of Fisheries. In this capacity he carried on fishery investigations from 1871 to 1874. Some of his articles were communicated to the journal of Asiatic Society of Bengal. Of these the longest and most important is the monograph on Cyprinidae which was printed in six parts.

A report on the fish and fisheries of fresh water of India was also published in 1871. However, a complete report on the fresh water fish and fisheries of India and Burma was published in 1873. He published his 'Fishes of India' in parts between 1875 and 1878. The book is considered to be a fairly complete account of the fishes of the Indian empire including Burma and Ceylon. Besides this he also contributed a large number of papers to the Zoological Society of London. The results of his studies on the classification and description of Indian fishes were also published in two volumes under the title "Fauna of British India, Burma and Ceylon" in the year 1889. His investigation convered a period of nearly 20 years. It contained valuable description of the rivers and irrigation tanks and the fish fauna including their migrations and breedings habits. He identified and described 351 genera and 1418 speceis of fish.

The excellent work done by late Dr. Day was of a very extensive nature. It maked the beginning of a new era in the history of fishery investigations in India. Had the investigations of Dr. Day been continued and had the various recommendations which were elaborated as a result of researchers in his life time been given effect to, there can be no room for doubt that a very great and marked improvement in the Inland fisheries of India would have resulted. Unfortunately, after the death of Dr. Day in the year 1889, the enquiries ceased and so far as the Inland fisheries of India is concerned, no futher work was done on the fisheries until 1906.

Capt. R. Beavan (1877)

The Hand Book of the fresh water fishes of India' was published in 1877 by Capt. R. Beaven. In his book, he has described the mischievous habits of poisoning streams and also of many species of fishes which grow to large size in tanks. He has also given hints on the stocking of fish in a pond.

Dr. T.C. Jerdon (1835–1897)

The fresh fisheries of Southern India was investigated by Dr. T.C. Jerdon who joined as a member of Medical Service in Madras in 1835 and spent sixteen yeas in various parts of the Madras Presidency. During the next five or six years he visited Punjab, Kashmir and the hill stations of great northern range. He published a report dealing with fisheris of Southern India in 1894. In 1897, Dr. Jerdon remarked on necessity for adopting some restrictive measures for preventing the destruction of large fresh water fishes such as the Mahseer during the spawning season.

For the first time in India, the attention of Indian students in Madras (1857) and Punjab (1882) Universities were drawn towards the fisheries studies. A number of surveys and expeditions carried out during British regime in India and adjoining areas have added much information on Indian fisheries, For example, the Yannan expeditions (1868 to 1875), the Persian boundary commission (1870–1872), the second Yarkand Commission (1873–1874), the Dafla expedition (1874–1875), the Afghan-Delimitation Commission (1885), the Afghan-Baluch Boundary Commission (1896), the Pamir Boundary Commission (1896), the military expedition of Lahasa in Tibet (1903–1905) and the Abor expedition (1911–1912), have not only brought to light some rare groups of animals and fishes but also the peculiarities of distribution of fishes of the Indian subcontinent.

K.G. Gupta (1906–1908)

On 26th July 1906 Sri K.G. Gupta a senior Member of Board of Revenue, Bengal was placed on special duty by the Government of Bengal in order to enquire into the fisheries of this province. His detailed and careful enquiries brought out the fact that no less than 80% of the population of Bengal eat fish. He further noted that whilst people do not eat fish every day but it may be safely estimated that for at least 320 days of the year fish form a very important part of the diet of the people in this province.

Dr. J. Travis Jenkins (1908–1909)

Dr. Jenkins made a preliminary survey of the Estuarine fisheries in the Gangetic delta and found that the fish fauna varied greatly according to the degree of salinity and the depth of water. He observed that net suitable for one part of the estuary might be unsuitable for other parts. He also stated that the fisheies offered a profitable return on any capital that may be invested on a properly organized scheme.

Dr. N. Annandale (1907)

Annandale in 1907 started two research publication namely Zoological Journals Records and Memories of Indian Museum which materially advanced the cause of fisheries research in the country. Dr. Annadale as a Superintendent of the Indian Museum rendered a great service through out the enquiry made on the fisheries of Bengal by H.C. Wilson, Dr. Jenkins and K.C. De.

In 1916 the Zoological and Anthropological section of the Indian Museum was converted into Zoological Surey of India.

Mr. T. Southwell (1911–1918)

In 1911, Mr. T. Southwell was appointed as the Deputy Director Fisheries. In 1917 Fisheries Department was again separated with Mr. Southwell as its Director. During the year 1911, the first attempt was made in Bihar to artificially propagate Hillsa (*Clupea ilisha*) on the lines adopted for shad (*Alosa sapidissima)*. T. Southwell published his report on Hilsa hatching operations conducted at Monghyr during August and October in 1912. A report on Madras fisheries Department was published by James Hornel in 1923. In the report, an account of active works regarding fishery development was given.

The morphology of Indian fishes was studied by a number of workers in India. The skelton of fishes was described by Sarabhai (1933) and Dharmarjan (1936). Bhimachar (1934) described the cranical osteology of *Ophicephalus straitus.* Awati and Bal (1933–1934) studied the endoskelton and blood vascular system of globe fish, *Tetraodon oblongus.* Awati and Pinto (1937) described the skelton of *Harpodon nehereus.* Sen (1926) gave an account of the branchiocephalic blood vessels of the fresh water carp, *Labeo rohita.* Roy and Das (1931) described the portal circulation in the siluroid fish, Arius. Mitra and Ghosh (1931) studied the internal anatomy of certain species of the families Chaudhuriidae and Mastacembelidae and Rahimullah (1935) the structure and function of the pyloric caeca of Ophicephaliadae and Notopteridae. Nair (1937) studied the changes in the internal sturcutre and the development of the bladder of *Pangasius.* A short paper of Mokkerji (1934) on the development of the intervertebral ligament in some fishes is also of interest.

A stimulus to work on fish and fisheries was provided by the general improvement in the facilities available for biological reseach at the sciene laboratories of the universities and the schemes of research subsidized at Calcutta, Madras and Bombay by the Indian (formerly imperial) Council of Agricultural Research. The Zoological Survey of India continued to maintain the lead on work on taxonomy and zoogeographical distribution of fishes.

The following is the list of the schemes subsidized by the Indian Council of Agricultual Research:

1. Life-history, bionomics and development of freshwater fishes in Bengal-Calcutta University, 1936–46.
2. Rural pisciculture-Madras Province, 1942–51.
3. Processing of Fish-Government of Baroda, 1943–45.
4. Improvement of preserved Fish Industry in Bengal, 1944–46.
5. Fish Eggs and Larvae of Madras plankton-Madras University, 1944–47.
6. Fish Eggs and Larvae of Bombay waters-Bombay University, 1944–47.
7. Bionomics of Indian Migratory Fishes- Bengal Government, 1946–48.
8. Experimentationon the Barn type of dehydrationp plants for preservation of fish-Orissa Government, 1945–1947.
9. Award of scholarships for fishery research-Calcutta University.

As a result of the study of various collections of fishes made form different regions of India, substantial progress in our knowledge of the fresh water fish fauna was achieved. The contributes (region-wise) are listed in the table given below:

Fish Fauna of India Studied During the Period (1938–1950)

Panjab

Lahore .... Ahmad (1943)

Uttar Pradesh

Kumaon-Himalays .... Menon (1949)

Rihand .... Hora (1949)

West Bengal

Kalimpong .... Hora and Gupta (1941)

Bihar

Hazaribagh .... Das (1939)

Rajmahal .... Hora (1938)

Parasnath Hills .... Menon (1950)

Hyderabad .... Rahimullah (1943–1944)

.... Mohmood & Rahimullah (1947)

Patna State .... Chauhan (1947)

Madhya Pradesh

Raipur District .... Hora (1940)

Satpura (Hoshangabad) .... Hora and Nair (1941)

Bastar State .... Hora (1938b)

Bombay

Bombay & Salsette .... Kulkarni (1947)

Poona .... Fraser (1942a & b)

. ... Hora & Misra (1942)

. ... Suter (1944)

Deolali .... Hora & Misra (1938)

Mysore .... Bhimachar & Rau (1941)

.... Hora (1942)

Travancore–Cohin .... Hora & Law (1941b)

. ... Silsa (1949)

Andamans .... Hora (1940a; 1941)

. ... (1940)

Carp Breeding

The major carps of India are known to spawn in June August when the rivers are in floods. Except in bundhs adjoining the river into which the flood-waters spread, they do not breed in confined waters. Hamid Khan (1942) put forward the view that the carps spawn when the rivers and streams are flooded by monsoon rains inundating the fields. In the symoposium on spawning of carps organized by the National Institute of Sciences in 1945, various workers in the field took part. Hora (1945) gave a historical account of the principal problems of carp spawning. Mookerjee (1945) laid emphases on the oxygen content of water as the principal controlling factor in spawning. Although the rains, water temperature and other physico-chemical factors apparently influence spawning, Hamid Khan (1945) emphasized the possible role of internal secretions which might be brought about by changes in the meterological conditions. Hussain (1945) formulated that the first requisite for carp spawning was flood which stimulated the fish to come out and indulge in sexual play, the spawning taking place if the temperature continued to remain below 96°F. Das and Das Gupta (1945) and Mokerjee (1945b) gave accounts of the breeding of the principal Bengal carps. Mokerjee, Mazumdar and Das Gupta (1944) paid special attention to the reproductive habits and collection of eggs and fry of carps of the Midnapur District. Ganpati and Alikunhi (1948 & 1950 described the off season spawning of carps and concluded that the availability of shallow grounds is the deciding factor in spawning.

One of the main difficulties in carp culture is the separation of the fry of the culturable species from those of predators. A beginning in this direction was made by Mokerjee, Mazumdar and Das Gupta (1944) who described the spawning habits and early development of *Labeo gonius* (Ham.). The life history of *Labeo gonius* was dealt with in detail by Moderjee and Ganguli (1949). They found regular segmentation of the eggs as found in the case of *Cirrhina mrigala* and *Labeo calbasu*. Jones (1938) described the external features in the development of the Ceylonese mountain carp, *Garra ceylonesis*. Khan (1943) reported the breeding habits and development of *Cirrhina mrigala*.

Chako and Kuriyan (1948) found *Catla catla* to breed in Indian water in the month of June.

The Mahseer proper (*Barbus tor*) is a well known migratory form whose movements are probably controlled by the temperature of the water. On the basis of maturation of the gonads, Khan (1939) found that there were three well marked spawning seasons in a year *i.e.*, in January-February, in May-June when the snow melts and in July-September during the monsoon. The breeding in May–June was not confirmed by Codrington (1946) who made a similar study.

The breeding of the Katli, *Barbus* (Lissochelus) hexagonolepis was dealth with by Langdale Smith (1944), Ahmed (1946 & 1948a) and Alikunhi (1948). They conclude that the fish breeds from April to October with a peak period in August and September. Ahmed (1948) found that the fish could be induced to breed in tanks.

Breeding of Other Fishes

Apart form carps, the murrel (*Ophicephalus striatus*) which is esteemed as food both in South and Central India was found to spawn when about two years old (Jones, 1946a,b). In Hyderabad, the, breeding season lasts from May to August, the larvae hatching within 24–48 hours (Hussaini & Rahimullah, 1946; Bhattacharya, 1946).

The nesting habit of the goramy, *Osphronemus goramy* varies as pointed out by Amirthalingam (1939) and Jones (1939). A detailed and careful study by Kulkarni (1943) gave an account of the breeding, nest building, larval behaviour and the methods of rearing the fish in aquaria. The detailed account of its nest building as outlined by Bhimachar, David and Muniappa (1944) confirms Kulkarni's (1943) observations. Job (1941) described the eggs and developmental stages of the spiny eel, *mastacembelus pancalus* (Ham) which is reported to attain sexual maturity in a year.

The spawning habit and development of *Wallagonia attu*, an active predator in fish-ponds and rivers was described by Ahmed (1944). The catfish, *Clarias batrachus* (Linn.), which contributes to a valuable fishery in Bengal, was investigated by Mokerjee & Majumdar (1950). These authors described the bionomics, nest building habit, sexual dimorphism, breeding and development of the fish. The unusual breeding habits of some of the catfishes of the genus *Macrones* were briefly reported by Raj (1940). According to Job (1940) *Aplocheilus punchax*, is a perennial breeder with a peak spawning period in June-August. Jones and Job (1940) described the life histories and larval stages of *A. lineatus* and *A. panchax*. Jones (1944) indicated the possible occurrence of physiological diapause in the eggs of Indian Cyprinodonts. Mokerjee and Basu (1994) dealt with the various aspects of Cyprinoudont life-histories. The contribution by Kulkarni (1940) in the life history and spawning of the remarkable fish, *Horaichthys* is noteworthy. The breeding habit and early development of the air-breathing Anabantid, *Macropodus cupanus* was described by Jones (1940) who drew attention to the occurrence of cement glands in the larvae. Jones (1946a, b) also gave a comprehensive account of the breeding and development of Indian fresh water and brackish water fishes based on published literature as well as on his own observations.

Breeding of Hilsa

The first step in the investigation on the bionomics of *Hilsa* was the discovery of the post larvae and young stages in the settling tanks of the water works at Pulta where water is pumped in from the neighbouring Hooghly River. It was inferred from this finding that the fish spawns in the Hooghly in the neigbourhood of the water works (Hora, 1938). Nair (1939) describd the early stages of development in considerable detail based on material obtained from the Pulta water works. Continued observation of Hora and Nair (1940) confirmed the earlier findings that spawning occurred in the Hooghly for the greater part of the year, with a main peak in July–August correlated with flooding of the river and a secondary peak in May. Evidences were available to show that the growth of the larvae was fairly rapid during the monsoon period but slowed down during the winter months. The number of young *Hilsa* in the Hooghly was found by Jones and Menon (1950) who obtained fertilized eggs and embryonic stages from the river in both sub-surface and bottom hauls in the vicinity of Pulta.

Migration of Hilsa

Prasad, Hora and Nair (1940) in their study of the sea ward migration of *Hilsa*, brought evidence to show that after leaving the river, the fish do not go far into the sea but move about in shoals in the esturaries extending from Bengal to the Orissa coast where *Hilsa* is a known to constitute a valuable sea fishery.

Hilsa Fisheries in Madras and Bombay

Chacko and Ganapati (1949) and Chacko and Krishnamurthy (1950) made observations on the *Hilsa* fisheries of the Godavari river where the fish ascend in shoals during the monsoon. Young fish reside in the tidal zone of the Godawari river throughout the year. The shoals of *Hilsa* were found by Devanesan (1942) to occur in the sea along the Eaast coast. Kulkarni (1950) gave in interesting account of the *Hilsa* fisheries of the Narmada river on the West coast. The bionomics of the fish on both the East and West coasts are more or less the same except for the range of migration limited to a distance of nearly 80 miles on the West coast, a plausible explanation for the limited movements being the steeper flow of the West coast river systems which prevents the more extensive migration observed in the Ganga system. Variations in the intensity of the *Hilsa* fishery according to lunar phases were noticed in the Balasore fisheries by Majumdar (1939).

Observations on the scales of *Hilsa* with a view to finding growth markings were made by Chacko and his co-workers (Chacko, Zobairi and Krishnamurthy, 1949; Chacko and Krishnamurthy, 1950) but the data gathered did not lead to any definite conclusion on its growth.

Hilsa kanagurta, which is closely related to *Hilsa ilisha* and occurs along with latter in Pulicat and Collair lakes and in the Godawari, Krishna and Coleroon rivers, breeds in the tidal zone from August to November.

Hilsa toil which is a marine species, forms a commercial fishery along many parts of the East coast of India from Vishakhapatnam to Tuticorin. It also forms a valuable fishery in the Narmada and the Combay coast, especially at Kodinar (Pillay, 1948).

Fisheries Administration and Management

The oldest established department of Fisheries in India is in the State of Madras, where with the efforts of late Sir Frederick Nicholson, the problems of fisheries received much attention leding to the formation of the Department in 1905.

When the Indian Constitution of 1935 came into force Fisheries was considered as a provincial subject in the same manner as Agriculture. While some states like Madras and Bengal forged ahead and implemented several schemes for the development of fisheries Several other states although rich in potential resoures hardly paid any attention to the development of fisheies. The reason for this apathy was that the influential and progressive communities in these provinces were generally not inclined to accept fish as an important item in their diet. As a result of increasing food shortage in the country brought poignantly before the public mind by the great famine of Bengal in 1943, the development of fisheries on an all India basis was felt by the Central and State Governments. The Government of India decided that some central agency was necessary to advise and co-ordinate development in the various sectors of fisheries. A section of fisheries development attached to the Ministry of Agriculture was created.

The Advisory board of the Indian Council of Agriculture Research in its well-known Memorandum of 1944 (generally known as the Kharegat Memorandum after the then Vice-Chairman of the ICAR laid down some of the essential directions in which fisheries development was to be achieved in the country. Among these were (1) the establishment of the Central Fish Committee and of a Fisheries Research Station, (2) the starting of pilot projects for mechanization, catching and storing fishes, (3) the development of pond culture practices and (4) the improvement of fish transport. Another official document of great interest is the report of the Fish sub-committee of the Policy Committee on Agriculture, Forests and Fisheies (1945) which recommended to the Government to embark on a comprehensive programme of work based on (1) survey of fishing area, (2) initiation and co-ordiantion of research, (3) conservation, development and exploitation of fisheries, (4) improvement of the socio-economic position of the fishermen (5) provision of more efficient and modern craft and tackle and, (6) the organization of the fish trade on proper lines. The importance of the development of fisheries was also indicated in the report of the Bengal Famine Commission (1945) as well as in the report on Scientific Research in India by Hill (1945).

Two official documents, the first by Hornell (1947) suggesting methods for developing marine fisheries and the second containing the memorandum on the proposed Fisheries Research Institute by Swell (1947) deserve special mention. The latter forms the basis on which the two Fisheries Research Stations of the Central Government were established.

Among the various reports of the State Governments on the development of fisheries, the best available accounts are those of the states of Bengal, Madras and Bombay. The official report on a survey of the Fishereis of Bengal by Naidu (1939) is a comprehensive account of the fisheries of the area. The subsequent annual reports of the Director of Fisheries, the Proceedings of the Conference of the Fisheries Officers and the Bengal Fisheries Development Pamphlets contain much material of interest. Popular accounts of the activities of the fisheries were published by several specialists of the Madras Government. The Government of Travancore put forward certain suggestions for increasing the quantity and quality of fish species under existing conditions in a Report drawn up by John (1945). Details of progress in work carried out in Travancore are also found in the Report of the Department of Research issued by the University of Travancore. Fishery development of Hyderabad was dealt with by Rahimullah (1944) and Hussaini (1949). A survey of the fisheries of Mysore was made by Bhimachar (1942). Official reports issued by the Department of Industries and subsequently the Department of Fishereis, Bombay, and the various reports and bulletins issued by the Department of Fisheries, Baroda, outline the activities in the respective areas. Of the varius surveys carried out special mentions may be made of the fisheries of Bombay (Setna, 1939; Setna and Kulkarni, 1946–1947), of Baroda (Moses, 1942), of Patna (Chauhan, 1947) and the fisheries of the Indus (Khan, 1946–1947).

Closely related to fisheries development and research, the urgent need for the setting of Biological stations in India was emphasized by Rao in his Presidential address to the Indian Science Congress in 1942. Several papers and articles dealing with the need for developing fisheries on scientific lines were published by Hora and others. The main emphasis was laid on the development of pond culture and the improvement in the fish cultural practices.

Transplantation of Fishes During British Period

Though transplantation of food fishes from their natural habit to nurseries and rearing ponds has been in vogue in India for a very long time. The augmentation the fish fauna by introduction of exotic forms and intrazonal transplantation of suitable autochthonous species for permanent establishment is of comparatively recent origin. The Madras Fishereis Department was, probably, the first government organization to take up transplantation of fish. The fishes that have been introduced or transplanted can be grouped as follows: (1) food fishes (2) game fishes (3) Larvicidal fishes and (4) ornamental fishes.

Transplantation of Food Fishes

One of the earliest recordings of transplanation of fish in India, is that of the Milk fish *Chanos chanos* during the latter part of the 18th century by Hyder Ali of Mysore and in Coondapur estuary in south Kanara by Thomas (1870). An interesting attempt to transplant *Hilsa ilisha* was made by Nicholson (1915). The ova of this fish were transferred from the hatchery to the Ponnai river in Malabar but there is no evidence of the fish having established itself there.

The catla (*Catla catla*) is one of the major carps of India and is reputed to be one of the fastest growing fish species. The successful transplantation of this fish in the south was achieved by the Madras Fisheries Department. The fingerlings from the Godavari river were introduced into Gunddapah Kurnool canal where they established themselves and spread into the Pennar river and the connected tanks in the Nellore Distirct (Raj. 1916). Catla fingerlings from the Godavari were introduced in the year 1921 into the Cauvery river below the Hogaikanal falls and into the Bhavani (Hornell, 1924) where the fish now affords a major fishery. Catla fingerlings were sent to Cochin by Job in 1945 where they thrived well. Fry from the Godawari river were also introduced into the Periyar lake in Travancore–Cochin (Chacko, 1948). The Bombay Fishereis Department transplanted catla from Patna (Bihar) into the Powai lake in Bombay where it bred and established itself (Kulkarni, 1947).

The Rohu, *Labeo rohita* is another major carp of India and the most esteemed fish in Bengal and Bihar. The earliest attempt to transplant this fish was made when fingerlings were taken from Calcutta and introduced into the fresh waters of the Andamans. From the records of Annandale and Hora (1925) and Mokarjee (1935) the fish grew well but it is doubtful whether it bred there. The Madras Fisheries Department had been regularly stocking this fish in the State from 1944 to 1949 with fry obtained from Bengal and Orissa (Jaganatham, 1946; Thyagarajan & Chacko, 1950) and attempts wre made to transplant the fish I the river Cauvery also. In Bombay, fry from Patna (Bihar) were introduced into the Pawai lake along with *L. calbasu* (Ham.) where both wre reported to have bred (Kulkarni, 1947).

The Mrigal, *Cirhina mrigala* (Ham.) is another important major carp, which is distributed throughout the North-west provinces of India. Its fry from Bengal were introduced regularly from 1943 to 1947 into the waters of Madras including the Cauvery (Thyagarajan & Chacko, 1950). Mrigal introduced as fry into the Powai lake in Bombay from Patna (Bihar) were reported to have bred there (Kulkarni, 1947).

The gourami, *Osphronemus goramy* (Lacep) fish, a native of Indonesia, was first introduced in India during the early half of the last century and stocked in the Botanical Garden at Calcutta, but the entire lot perished by 1841 for want of proper attention (Thomas, 1881). About the year 1865, Sir William Dension, the then Governor of Madras, imported some gourami from Mauritius and introduced them in the Government house ponds at Madras while some were taken to the Nilgiris (Raj. 1916). The fish established itself in various parts of Madras State from where it was successfully transplanted to Bombay in 1937 (Kulkarni, 1943), Baroda in1941 (Mosses, 1944), Mysore in 1941 (Bhimachar *et al.*, 1944) and Cochin and Hyderabad in 1945. Gurami was introduced in the Punjab but could not survive to low winter temperature (Khan, 1947).

The Tench *Tinca tinca* (bloch) was brought from England by Mr. Maclvor in the year 1870 along with the golden carp and was introduced into the Ootacammund lake (Molesworth and Bryant, 1921). The fish bred in the lake and subsequently fingerlings were transplanted to some more ponds and lakes in the Nilgris and the Shevaroy hills.

The Crucian carp, *Carassius carassius* (Linn) also known as golden carp, is a native of Central Europe from where it has been transplanted to various countries. Maclvor introduced it about the year 1870 along with the tench into the Ootacmund lake where it bred well (Molesworth and Bryant, 1921). Subsequently it was transplanted to several ponds and lakes in the Nilgiris, Shevaroys and Kodaikanal. Later on established successfully in Ceylone and is said to be in abundance there. From there a consignment of 45 young fish was brought by Dr. Sundara Raj, the then Director of Fisheries, Madras in 1939. The mirror carps were stocked in the Ootacamund lake where they thrived well and bred. In 1946 some fingerlings of the mirror carp were introduced in the Ulsoor lake, Bangalore (Burton, 1948). In 1947 few fingerlings of this fish were transported by plane in an oxygenated containers from Ootacamund (Nilgiris) to Bhowali in Kumaon Hills (U.P.) (Raj & Cornelius, 1948) and the fish bred there.

Transplanation of Game Fishes

The Mahseers *Barbus* (Tor) *putitora* (Ham.) are the favourite among Indian game fishes. The putitor mahseer is found all along the Himalayas from Kashmir to the Darjeeling hills. The lakes of

Kumaon hills were stocked with this fish by Sir H. Ramsay in the year 1858 (Walker, 1988). The Bhimtal, the Nakuchilatal and the Sathal were stocked with fingerlings transported in earthen vessels from the Gola river and the Koli river. The fish took well to the confined waters and bred in the shallow areas of the streams that drain into the lakes.

The khudree mahseer, *Barbus* (Tor) *khudree* (Sykes): (Hornell, 1923) was reported to have been stocked in the Kodaikanal waters (Palni Hills) with 162 mahseer fingerlings from the Tungabhadra in Kurnool.

Trout is the only exotic game fish introduced into India and at present two species have established themselves, *viz.* The Rainbow trout, *Salmo gairdnerii* (Richand) and the Brown trout, *Salmo trutta fario*, the former in the south and the latter in the north. The first attempt to introduce trout in India was made by Mr. H.S. Thomas in 1863, but the consignment of ova perished on the way (Day, 1876). In 1866 Day imported 6,000 ova and most of them died a few days after reaching Ootacamund, the few that survived turned out to be the first trout to see life in Indian waters. The credit for the permanent establishment of the trout in the Nilgiris goes to Mr. Wilson who organized the whole work on scientific lines. He constructed a hatchery at Avalanche and concentrated on the establishment of the rainbow trout. All the available stocks of brood fishes were transferred and fresh consignments of ova were brought from Germany and New Zealand, and fingerlings from Ceylon. The Avalanche hatchery was a great success and the rainbow trout is now well established in the Nilgiris waters.

The introduction of Brown trout in Kashmir was carried out independently to that at the Nilgiris. The first shipment of ova was sent in the spring of 1900 as a present from the Duke of Bedford to the Maharaja of Kashmir in return for the Kashmir stages presented by the latter (Mitchell, 1918). Brown trout and Rainbow trout were introduced in the snow-fed Himalayan streams in Kulu and Kangra by Mr. C.C.L. Howell, the first Director of Fisheries (Punjab) in 1912. These fishes have spawned ever since 1912 and the steams at Kulu, Kangra, Chamba and Mandi have been stocked form their spawn.

Introduction of trout in the Kumaon lakes and in the Eastern Himalayas was brought in 1910 when a consigment of 10,000 ova was taken from Kashmir to the Bhowali hatchery, about 8 miles from Nainital. Another consignment was obtained in 1912 and fingerlings from these were stocked on varius lakes such as the Nainital, Naguchiatal, Sathtal, Malwatal and others (Skene-Dhu, 1923).

Introduction of brown trout in Travancore Kodaikanal and other waters in South India was obtained during the period 1909–1913. By this time the culture of the Rainbow trout was becoming a notable success in the Nilgiris and attention was turned to this species. Though the fry introduced into the waters of the High Range showed phenomenal growth, they failed to breed until a hatchery was located at Rajmally and the fish were liberated in the Eravikulam river, whee they bred under natural conditions in 1937. Fresh stocks of Rainbow trout were obtained form the Nilgiris and Ceylon and by 1941–1942 the fish had firmly established itself in the high Range of Travancore (Gopinath, 1942 and Mackay, 1945).

Transplantation of Larvicidal Fishes

The Top Minnow, *Gambusia affinis* from North America has been introduced in varius countries for larvicidal purpose. It was first introduced in India by Dr. B.S.Rao and Dr. Chandresekhriah who brought an experimental consignment to Mysore from Italy in 1928 (Gopinath, 1942).

The attempt was a great success and the fish bred out of this stock was transplanted from Travancore in the south to the Punjab in the north. A consignment of young *Gambusia* was brought from Ceylon in 1929 by the Madras Fisheries Department and stocked in the waters of Krusadai Island, and another was taken form Banglore to Madras city in 1930 (Chacko, 1949).

The Millions, *Lebistes reticulates*, is a native of South America and is known as 'Bnarbadas Millions'. A consigned was introduced by Major Salley into Madras in 1909, but was reported to have perished due to non adjustment of climatic conditions (Prashad and Hora, 1936). However in 1946, the fish was noticed thriving in the Rameshwaram temple tank and from there it was successfully transplanted to various parts of Madras (Chacho, 1948).

Transplantation of Ornamental Fishes

Records are not available that can show the exact period of introduction of ornamental fishes into the country. The commonest and the most popular ornamental fish is the gold fish, *Crassius auratus* which was brought from China. Other examples are the angel-fish, *Pterophyllum scalare* from South America, the fighting fish, *Betta pugnax* from Siam, the white cloud mountain Minnow, *Tanichthys albonubes* from China, the black tetra, *Gymnocorymbu ternelzi* from Paraguay, the flame tetra, *Hyphessobrycon flammeus* from Rio de Jeneiro, the pearl Gourami, *Trichogaster leeri* from Siam, the jewel fish, *Hemichromis bimaculatus* from Africa, the Chinese paradise fish, *Macropodus opercularis* form China, the red and green Swordtail, *Xiphophorus hellerii* from Mexico, the different colour varieties of Platies, *Platypoecilus maculates* from Maxico, the liberty Mollies, *Mollicenisia sphenops* from Texas, the Cherry Barb, *Barbus titteya* from Ceylon, the Negro Barb, *Barbus nigrofaciatus* from Ceylon and the Harlequin, *Rasbora heteromorpha* from Sumatra and Singapore. Exact details about the introduction of these ornamental fishes are not available.

Accidental Transplanations

In the course of intentionare introduction of species accidental introductions also took place. The case is that of *Rosbora daniconius* (Ham.) and *Oryzias melastigma* (McCl.) into the Andamana along with the fry of *Labeo rohita* (Mokerjee, 1935 and Hora, 1941). *Ophicephalus gachua* (Ham) found in the Andamans is considered to have been introduced accidentally by human agency. According to Annadale and Hora (1925) the fry of minor carps and certain Siluridae might have been introduced into the island along with the fingerlings of *Labeo rohita*. The Tilapia, *Tilapia mossambica* (Peters) which is a native of South Africa, has given very encouraging results in some of the South-East Asian countries where it has been introduced in recent years. The usefulness of *Trichogaster pectoralis* (Regan) for introduction in paddy-fields, swamps and marshes requires to be studied. This is a native of Siam and is now well established in several parts of Malayasia and Indonesia.

Fishing in the British Periods

Day (1875–1889), Thomas (1873), De (1910), Hornell (1942) and Hora (1926a,b) described the common and extremely diversified methods employed in fishing in Inland water of India. Thomas (1873) has given some remarks regarding Indian fishing in his book "The Rod in India". De (1910) in his report on the fisheries of Eastern

Bengal and Assam described following methods which are employed for the capture of fish:

Nets

Seine-nets, Pocket-seines, drag nets, drift nets, caste-nets stake-nets, fixed purse nets.

Traps

Manipulated traps, fixed traps

Spears

Baited hooks and springs: Fishing rods, bottom lines, baited springs

In "The fishing methods of the Ganges" Hornell (1942) classified the Gangetic fishing methods into four categories:

1. Hunting
2. Netting
3. Angling
4. Miscellaneous

Hunting

The compound spears made up of numerous simple lances firmly bunched together towards the butts but divergent and free for more than half of their length are much in evidence throughout the low-country. It is known generally as Konch. The simple spear or the harpoon is more common. Invariably the harpoon is barbed either single or double. The shaft consists of a bamboo, 9 to 10 feet long. The barbed point, 9 inches or so in length, is usually fitted to a short wooden handle and this in turn into the hollow end of the bamboo shaft. A cord connects the wooden base of the harpoon point with the shaft.

The konch or compound spear has also its harpoon form, when it is knows as juti or jutiya. It is used in the same way as its prototype but having barbs on the small iron spear heads, which are attached by thin cords to the long conical 'handle'. The prey is more security held when struck. If the fish is large and puts up a hard fight in its attempt to escape, the great basal cone serves as a buoy easily kept in view by the pursuing boat. A variation of this compound harpoon is the pacha of Patna.

Spearing fish by torchlight is frequently practiced. The use of bows and arrows is extremely effective in the hands of expert archers. It can be practiced only in perfectly clear and transparent waters. As it requires lengthy practice to attain proficiency, it is confined to a few particularly favourable localities. The blow-gun used commony to shoot fish with a harpoon dart in the northern parts of Travancore and the adjoining taluqs of Malabar is not known in the Gangetic valley.

Trapping

The most primitive of all is the separtion of some shallow area of water from the main body by the simple method of throwing up a low earthern bund between the two and then by bailing the water out. The fishes present into it are let behind stranded. Another primitive form of trapping fish is to laydown bushes and branches in water to attract fishes to the shelter. In Bengal small tanks and pits are found everywhere in the country around the ryots' homesteads. During the rains when they are full of water branches are thrown into them. As the water level goes down at the end of the rainy season, large number of fishes are trapped into these pits (khatis).

Boats not in use are kept submerged to prevent damage to the wood. The interior is often filled with leafy branches to attract fishes. When a net is thrown over the boat and is drawn ashore, the sheltering fishes are caught.

The only snare commonly employed is the baited spring, called Barra, Kai-barsa and Datia (Rangpur). It is st free in inundated paddy fields, the favourite resort of the climbing perch (Koi).

Fishing by Hand

Fishing with the help of hand is the simplest and most primitive method of fishing, especially at the riverbanks and the sides of the boulder. Usually a truncate cone open at each end is used as the hand trap. A typical polo or Tappu, as it is generally termed in Bengal, varies from 2 to 2½ feet in height. The lower aperture, which is made as wide as can be conveniently operated has a diameter of about 24 inches. The upper aperture is made just sufficient to permit the hand and arm to enter comfortably into it. The frame of the ordinary Gangetic form is made of split bamboo supported internally by a number of bamboo hoops. The Polo is used only in shallow water, not exceeding two feet in depth. The fisherman wades into the water, continually dropping the trap into the water and pressing down the wide mouth into the muddy bottom with his left hand. With the right, passed in through the narrow upper opening, he gropes around in the mud for any fish that the trap might have covered.

Several transitional forms connect the conical or bell-shaped polo with the scoop traps. These are the Chakjal and the Tura of Patna. The Chakjal is clearly a development of the polo. The Tura is used to catch muddwelling fish. It is operated like the polo, but as it is without an opening at the apex to admit the hand, it has to be up-ended with a scoop-like motion to lift out any fish caught.

The Chakjal leads into a blind alley. The other trap or net is derived form it, whereas the tura appears to be the parent of all the scoop trap in existence. It leads directly to the Nui, Beloin, or Ganj and thence to the bamboo form (honcha) of the lave-net.

The central form of the Nui Ganj is a cylindrical scoop one end brought to a closed eccentric apex at the other. A much larger form of scoop is the Honcha, which is essentially a primitive form of lave-net, where instead of meshes of thread or cord being employed to make the scoop or bag part, fine bamboo network is substituted. A still closer approximation to the lave-net is Hogra where the scoop is given a rigid mouth frame composed of a triangle of bamboo. Another device which utilizes the habit of certain fishes to seek shelter is holes and crannies in the Chunga. A number of these hungas are placed under water in holes in the bank of a tank or sluggish stream and left there for several days, to attract fish to make their homes in their enticing cavities.

Of cage traps the number and varieties are endless. The trap from which all these are derived is not a cage at all. It is merely a short cylinder of grass or fine bamboo matting that sieves out fry and prawns from some little funnel. The next stage in the evolution of this is the permanent attachment of the entrance funnel to the cylindrical boyd. Sometimes these traps attain considerable dimensions. Of such is the Paron, a truncate cylinder which may be five feet in length by three in height. It is used to catch large fish and is the largest kind of trap used in the Gangetic region. The Katya is another trap but with a smaller mesh (1½ inch). The Paor, a large sub-cylindrical trap of the Katya type is placed facing the current. Fish coming down stream enter the opening at the base of the funnel-shaped mouth and become prisoners. In the Dohair, a smaller trap, there is only one ladder stretching outwards from the center of the mouth. It is used near the banks of shallow streams to trap fish traveling in either direction.

Futher complication against egress is obtained by setting a second funnel-shaped trapped partition farming two chambers within the trap. The Kai-chai also known as Jalanga, is a good example of this form. It is usually provided with a cord carrying a handle on the upper side.

Nearly all the remaining types of cage may be classed as box traps being quadrangular in genera form. A transitional form is the Kholson, a cylindrical trap with truncate ends, set upright on the end. This useful trap is used in comparatively deep water with leafy branches on each side partly to attract fishes by their shelter and also to act as ladders. The Charo of Patna and Western Bengal is constructed on the same principal, but with the curves of kholson converted into rectangular corners.

Another set of more or less quadrangular traps made of finely split bamboos laced together very closely so as to leave very narrow interspaces are used to catch prawns and bottom-feeding fry and small fish. Of these the Boichna, Rabani, Antal, Doar and Darki are typical common forms. Bottle-shaped traps woven of fine bamboo wickerwork and cones of woven kusha grass with trapped mouth form another and separate minor carps. The small fish that enter are taken out through the neck opening which is closed when in use by either a little door or some form of plug.

One of the grey mullets, *Mugil corsula*, has become completely naturalized in the fresh waters of the Gangetic plain. The fish is extremely shy and very difficult to catch by ordinary methods. A floating raft made of mats margined by bright objects to induce the fish to jump and and upon the floating structure behind is employed to catch this fish. Of the several variations of this method of fishing, the commonest is the Chali or Chanchi, a rectangular float of reed matting, made from the nal reed, measuring from 20 to 30 feet long and 3 to 4 feet wide. The idea embodies in the use of these mats is also employed frequently in conjunction with the barring of small water courses, drainage channels, narrow creeks and the like with a view to catch fish both going upstream when the rains begin and on their return when they are over. A common plan is to throw up a bund or dam across the steam leading a narrow passage in the middle. This is barred by a strong bamboo screen which permits water to pass but not the fish. A boat or a raft is mored on that side of the screen opposite that from which the fish are approaching. Finding their passage barred, some species leap the barrier and in so doing fall into the waiting boat or upon the mat raft.

Netting

The people living along the banks of the river Ganges have not only displayed a great skill in desiging fish traps, but they have devised different types of nets as well. These include the following:

Sience Net

The sience net or the her-jal is probably the most extensively used class of net in the water-system of the Ganges. This net, is a huge stretch of stough netting of 1000 feet long and thirty feet in depth. It is shot from several boats and hauled ashore by a crowd of thirty to fifty men. The mesh size varies from a quarter of an inch to a full two inches, knot to knot. Gangetic seines have several notable characteristics such as the absence of sinkers, compensated for by the use of stout foot rope, the absence of bag or bunt at the mid-length of the net, and the general custom of tanning the net with gab. For floats a variety of materials are employed. Pieces of light woods, internodes of bamboo, bundles of pith and of reeds, are the most common. The larger seins, known as jagat-ber or inaha-jal are employed extensively in the largest estuaries and rivers. The smaller-sized seines are genrally called chooto-ber.

Drag-nets or Trawl-nets

The line of demarcation between seines and dragnets is often indeterminate, as many varieties of the seines are also used to drag shallow rivers, jheels and tanks. Conversely dragnets are frequently employed in the manner of seines. The Moi or Moia-jal, use din some form or other almost everywhere in Bengal and Bihar by two men, is one of the simplest of these. It is a small net used for dragging in shallow water. It is made of sunn-hemp, the meshes seldom more than a quarter of an inch, knot to knot. The usual size is 15 feet in length by 9 feet deep. About one third of the lower edge of the netting is doubled up on the front side, and sewn along the sides, forming a long pocket. A number of small drum shaped weights are attached along the lower side of the pocket-mouth. In the chunti-jal of Patna the lower margin is folder over to form a long pocket from end to end of the net.

Bag-Nets

Of the trawl form of bag-net, the toni-jal and the harchari-jal are typical of the most common varieties. The former is a plain rectangular bag as wide at the bottom as at the mouth. Head and foot ropes are present on either side of the mouth, which is kept open by the simple expedient of inserting a vertical bag at the mid-length of the opening. At the angles of the mouth are attached the two hauling ropes. In the kanti-jal, there may be a number of vertical distending rods across the mouth and the hind end of the bag may be divided into several pockets.

Stationary bag-nets are used only in estuaries and in wide and rather shallow rivers. The suti-jal is the simplest one. It consists essentially of a long tubular bag, with a mouth having a circumference smaller than that of the bag. From each side of the mouth extends a long leader of wide netting, 4 to 5 feet in height, supported on stakes and driven into the bottom. Another stake passes through the center of the mouth and besides helping to anchor the bag, serves to keep the mouth distended. The head and foot ropes are tied to it at a suitable distance a part. The orifice at the tail end of the bag is tied with cord when in use. This net is usually set in rivers with a strong current.

Special Nets Used to Catch Hilsa

Three specials net namely the Kharki or Sharki-jal, the Shangla-jal and the Kona-jal are used to catch Hilsa in the Ganges. The first two are purse-shaped nets with hinged mouths which can be shut instantly when desired; they may be termed clap-nets. The Kharki-jal is the simpler. It has a roughly rectangular bag of the same shape as that of the Toni-jal from which it appears to be directly derived. The Shangla-jal is a refinement of the Kharki-jal, more elegant in form and more delicate and responsive in manipulation. In outline the mouth is semicircular. The Kona-jal, also called Bhasha-gulli is one of the longest seines, measuring some 300 feet in length with a depth between the headline and foot rope of 30 feet. It is strongly made of hemp with a mesh of 2 ifches, knot to knot. Like all Gangetic nets it is tanned with gross.

Drift-nets and Gill-nets

These nets are derived probably from the seine and are chiefly used when the rivers are in flood. The most typical and powerful is the great Chhandi-jal which is employed chiefly to capture the *Hilsa* when the shoals are ascending the great rivers in the rainy season. The Chhandi-jal is made like all large drift nets in a number of short length pieces which are tied together into a long fleet. Each piece measures 36 feet in length by 24 feet in depth and twenty four pieces usually constitute a fleet. The nets are of hemp tanned with gab, and have a mesh of 1½ to 2 inches, knot to knot. Floats of bamboo or of light wood are used to buoy the head-line at intervals of about 20 feet. The foot rope is weighted at regular intervals usually with annular burnt clay sinkers, or fragments of brick. The net can be adjusted to float vertically at any desired depth. This large net is set in the evening from one or several of the specially broad boats called chhandi, built specially for this method of fishing.

Cast Nets

The ordinary cast net or khepla-jal of the Ganges is a conical net having an extremely wide mouth. The height of the cone varies from 15 to 18 feet and the diameter of the mouth is about 40 feet. The circular edge of the cone is doubled inwards to a height of one to two feet and sewn to the body wall of the cone at regular intervals, which forms a series of looped-up pockets. The free edge of each pocket is weighted with a line of closely set drum shaped iron sinkers, forming a continuous series all round the inner surface of the mouth of the net. A cord 20 feet long is attached to the apex, having a loop at its free end through, which the third and fourth fingers of the right hand are passed. The remainder of the cord and the whole of the net with the exception of a short pendant portionof the mouth end are piled in loops upon the bent right arm. With a quick dexterous motion, the body is swung to the left and the net given such an impetus that it whirls outwards from the fisherman's arms in a rapidly widening expanse. The net spreads to its full extent on the surface of the water. The weigths along the marginal pockets rapidly carry the periphery to the bottom, the unweighted apical region following slowly. The fishes within the area encircled by the weighted edge of the net are imprisoned. The cast-net in its different sizes is the commonest form of net used throughout the Gangetic valley.

Dip-nets

The dip-nets fall int two categories, those of triangular form and the others square in outline. The former are usually partially framed between two bamboos converging towards the fulcrum, the latter are usually suspended by the four corners from two crossed down curving arms. Triangular dip-nets seem to have been derived from the scoop trap. The simples form of the triangular dip net is a hand net operated exactly like a scoop, the broad anterior end being pushed along the bottom for a distance and then upended to throw any fish into the apical end close to the handle. This hand dip-net or the hela-jal is the commonst of all the many varieties in use. Larger editions of this type of net is the kharra-jal, a large triangular net worked from bamboo staging implanted in the river bed. The platform where the fisherman stands to work, the net is at a height of several feet above the surface of the water..

The rectangular dip-nets used on the banks of the Ganges form a complete series, from the little squqare of netting worked by a string in the hand, to the great nets many feet square worked at the end of a long beam and balanced by heavy counterpoise weights. Amongst the simplest is the Khola-jal or khorashula-jal, some 8 feet by 6 feet, of small mesh, stretched and suspended by the four corners from the four ends of two downwardly curved bamboos which cross at right angles in the middle end than the other. The net is palced in shallow water on the shelving margin of a river. A cord is tied to the place where the bamboos cross. This is specially used in catching 'Khorsula', a species of grey mullet living in fresh water.

In the dharma-jal, two slender split bamboo rods are either crossed mid-length and tied securely in this position or are passed through two short lengths of nal reeds, which are tied together crosswise in similar fashion. The free ends of the four arms are then tied to the four corners of a square piece of netting from two to three feet along each side. A short string from the point where the arms cross, is tied the end of a bamboo rod by which the net is dipped and raised. It is a very common fishing contrivance in Bengal.

Angling

According to James Hornell (1924) a great fascination for angling is most developed among the people of Bengal in comparison to the people in other parts of India. He described the methods of angling in his article 'Fishing methods of the Ganges'. Thomas in this classical work "The Rod in India" (1873) gave references as how to obtain sport fish with remarks on natural history of fishes. In the 'Tank angling in India' Thomas (1887) described stocking of ponds, the carriage of fish and the habitat of nearly 75 inland fishes worth taking with a rod. In the 'Tank fishing in India' Thomas (1887) described paste baiting, live bait picketing and live bait with a float.

The rod is usually home-made, cut from a carefully selected length of bamboo. Finer ones tapered more evenly come from the hills and find a palce in the local markets. In all cases, either the old fashioned tight line or the running line is used. In its simplest form the tight line is tied firmly to the tip of the rod. In such cases the rod when used for largest fish is short and strong. The landing of the fish is a matter of a quick strike and a dexterous throw out upon the bank. The rod is more slender and the tackle is light if the small fish to be caught. Ordinary rods used with tight and semi-tight liens vary in length from 5 or 6 feet even more. The lines are usually of well twisted cotton or hemp cord, tanned and stiffened by impression in gab decoction. Stronger and more durable fishing lines are made from tasar and refuse silk. Almost invariably the hooks are barbed. For surface fishing, a single hook is used; for bottom fishing two or three may be tied on. The indigenous hook is shorter in the shank with a wider and more angular bend than the typical European pattern.

Great varetiy characteristics the objects used as float. The favourite one is a length of the quill end of a peacock's feather. Others are made of short lengths cut from the flowering stems of reeds (nal), certain grasses (ulu) or the jute stem. Sections of sola-pitch are also frequently employed. The sinkers are now-a-days mostly made of lead strips wound round the line till the requisite weight is attained.

The ground baits comprise everything the angler can think of to tempt fish to his hook through the senses of sight, smell and taste. The favourite animal baits are earthworms, prawns and large silk-worms. Small live fish, particularly murrel fry and the little *Barbus stigma* are commonly used, together with a variety of beetles and frogs. Dough made of flour or of baked rice, fragments of jack-fruit and fish flesh are other useful baits. Perhaps the most curious bait is the cockroach used when fishing for the faint catfish, *Wallago attu*.

The value of ground baiting is so well understood that the manufacture and use of appropriate mixture have been brought to a high degree of specialization. They are particularly valued in fishing for several of the larger carps, notably *Catla catla, Labeo rohita, L. calbasu* and *Cirrhinus mrigala*. The mixture employed consists essentially of strong smelling substances bound up in clay or dough, in order that the diffusion of the odour shall proceed comparatively slowly. The resultant paste, made up into balls, may eithr be thrown into the water before or during fishing with a view to attract fishes to the spot.

Head-lines (tuggis) are also in common use and have several distinctive features. In nearly all cases, the line is wound round a perforated cylindrical roller, which revolves upon a short bamboo stake driven into the margin of the tank or stream. The roller in its simplest form is made from an internode of a medium-sized bamboo ornamented frequently with the marks. A tuggi and its roller are not complete without the addition of a little contrivance to notify the fisherman when a fish is hooked. A development of the hand line is the dhanbansi, a stout line used from a boat, having an array of baited hooks attached at short intervals.

Long-lines are employed extensively in all the large rivers. The common type is called chhara bari, hazari bari and also dogha (malda), and dhauni (Murashidabad). It is a stout line often of 100 feet or more in length, armed with a large number, 200 or even more, of short side lines or snoods, set at frequent regular intervals. To each snood a hook is tied at the free end. The main line, usually weighted with sinkers, and with all the hooks baited, has the two ends attached in such way that it remains in the same position till lifted.

Miscellaneous

Several forms of fishing with the aid of bushes of which we have seen the genesis in the filling of ponds and pits with leafy branches during the rains, are generally practiced throughout the whole of the low-lying deltaic plains of the Ganges.

Hora (1926b) described a peculiar fishing implement from Kangra Valley, Punjab. The specimen consists of three pieces of thin hemp twine, ech about a foot and a quarter in length. Of these two pieces are knotted, so that each forms a running noose and are firmly fastened to the two ends of the third central piece. To this piece are tied at irregular intervals a number of horse hair loops with running nooses. For this purpose a long hair is taken and tied to the twine in the center of its length, the two free halves are then knotted into loops. The implement is used to catch fishes in the rapid-flowing waters of shallow, rocky streams.

Fish Preservation and Processing in the British India

The methods of fish preservations employed in British India may be classified as follows:

1. Desiccation with or without salt.
2. The use of anti-septic preservation such as brine, vinegar etc.; Pickling and certain miscellaneous processes *e.g.* making spiced fish pastes.
3. Sterilization accompanied by hermetical packing (canning).

Sun Drying

The common variety of Sun-dried fish is called Sutki or Sukti. The drying is conducted in open yards called "Khola" large fishes are beheaded, gutted and the entire viscera removed. Very large varieties are cut into slices. The smaller fish are dried entire. The fish are laid flat side by side on mats made of nal-reeds and exposed to the Sun., Sometimes there are bamboo platforms 3–4 feet high to spread the fish. The drying proess is continuous and extends for a week or 10 days depending on the Sun shine. The fish are not removed at night but are protected from the depredations of birds and other animals by stretching a net over them.

The dried fish are preserved in earthen jars called Matkas, which are well smeared with fish oil on the inside. These jars are buried in shallow trenches in a shady place. The dried fish are also stored in trenches dug in the ground which are lined with straw and mats. After putting the fish the trenches are covered with earth. The common varieties of fish used for Sukti are Rohu, Calbasu, Catla, and sometime Catfishes too.

Curing

The methods of curing fish in India are sun drying without salt, salt curing by the dry process and salt-curing by the wet process. The object of curing is to withdraw natural moisture from the tissue of the fish to prevent bacterial actions as well as enzymatic decomposition, which act rapidly in the presence of moisture and warmth.

Curing with Salt

Only Hilsa fish is cured in this manner. The fish is cut open and eviscerated. The cavity is filled with salt and closed up. Salt is also rubbed over the fish. The prepared fish are placed in a pit with more salt and the pit kept closed.

Statement showing the amount of fish cured in India (1946) in thousand mounds:

Province/State	*Sun Dried*	*Salted*	*Total*
Bengal (Both Estuarine and seafish)	850	150	1,000
Assam	50	15	65
Other areas	21	03	24
Percentage of total catch of fresh water fish		0.3	1.4

Preservation in Mustard Oil

The fish is dried as for Skuti until it can be crumbled into a powder between the fingers. They are then collected in baskets and immersed in water for few hours. The baskets are then hung up for draining. The moist fish are then packed into large earthen jars called Matkas. The interior of the Matka is well soaked with fish oil. The fish are packed closely in side with a sprinkling of fish oil between the layers. The mouth of the jar is closed with a earthen ware. The jar is buried in the ground for several months to allow the fish to develop the taste.

Fish Pastes

In Jalpaiguri dried fish is powdered and mixed with the crushed stems of Caladium (kachu) and made into balls. In Nowgong, a fish paste called 'Hidla-khunda' is manufactured from certain varieties of small fish which are placed in a hole dug in the ground and kept covered for nearby a month. Some times a little ground pepper and some alkaline ash (called Khar in Assam) is mixed with the fish at this stage. In Cox's Bazar, a fish paste of prawn called "nga-pee" is prepared with 3:1 salt and the fluid drained off. The paste is basked and sold as "nag-pee". The Chittagong Bazar nga-pee is also called be-lee chong.

In the 33rd session of Indian Sciences Congress held in Banglore in 1946 Dr. B.C. Guha delivered his presidential address on the preservation technology. He attached great importance to the freezing techniquje in the preservation of fish. He remarked that quick-freezing was the first class method for preserving fish. He pointed out, however, that refrigeration did not save space in transport as dehydration did. Dehydration of fish, for instance, would reduce the actual consumable part of the fish to only about 1/20 of its original weight.

Marketing of Fish in the British Period

According to the report available on the marketing of fish in India, the following commercial fishes of Inland water were identified and the localities from where they were caught were also described.

Cat Fish Group

Wallago attu

Caught only in large northern rivers, small specimen seen in all fresh water.

Bagarius

Caught only in the northern rivers.

Clarias magur

Cught in rivers, jhils and tanks.

Pangasius buchanani

Caught in large northern rivers.

Silundi gangetica

Caught only in the large northern rivers.

The Herrings Groups

Clupea ilisha

Found in the estuarine region of Indus, Sabarmati, Cauveri, Krishna, Godavari, Mahanandi and the Ganges at the time of flood.

Bombay–Duck Group

Harpodon neherus

In the estuaries of Bengal.

Feather Back Group

Notoperus

Caught in the fresh and brackish water of India. The two species are known of which *Notopeterus chitala* is the bigger in size.

The Mackerel and the Perch Group

Cybium guittatum* and *Cybium commersonii

Are abundant near the estuaries of larger rivers.

Etroplus suratensis

In Brackish water and can be acclimatized in fresh waters.

Lates calcarifer

Are abundant near the estuaries of larger rivers.

Mullet Group

Mugil speigleri* and *Mugil oeur

Found in the estuaries of larger rivers.

Twenty five species of mullets are known, many of them ascend in rivers and can be acclimatized to residence in fresh water.

Mugil corsula

Found in the larger rivers.

Indian Salmon group

Polynemus tetradactylus

Caught chiefly near the mouth of rivers.

Polynemus paradisicus

Found only in the estuaries of Bengal.

Live Fishes Group

***Ophiocephalus* sp.**

Found in all tanks, pools, jhils mostly in stagnant water.

Anabas scandens

Found in both estuarine and fresh water.

Carp Group

Labeo* spp., *Catla buchanani*, *Cirrhina mrigala* and *Barbus tor

Found in steams, rivers and tanks all over India.

The table given below the estimated marketable surplus in the various provinces and states:

Province/State	Production (in Thousand of Manuds)	Percentage to Total Catch	Estimated Value (in lakh of rupees)	Percentage to total
British India				
Assam	721.6	11.5	46.22	6.23
Bengal	3133.2	50.1	432.92	58.32
Bihar	959.5	15.3	95.99	12.93
Bombay	81.0	1.3	9.82	1.32
Central Provinces	156.0	2.5	16.28	2.19
Coorg.	0.1	Neg.	0.02	Neg.
Delhi	5.6	0.1	0.80	0.11
Madras	187.0	3.0	13.96	1.88
North West Frontier	1.5	Neg	0.24	0.03
Orissa	326.0	5.2	34.92	4.70
Punjab	25.0	0.4	3.99	0.55
Sind	266.5	4.3	38.56	5.19
Total	**6010.0**	**96.0**	**716.49**	**96.52**
Indian States				
Baroda	9.8	0.2	1.99	0.27
Cochin	15.5	0.3	0.83	0.11
Gwaliar	8.1	0.1	1.25	0.17
Hyderabad	20.0	0.3	2.17	0.29
Jammu & Kashmir	17.8	0.3	2.88	0.39
Mysore	10.3	0.2	1.55	0.21
Travancore	74.8	1.2	4.01	0.54
Other States	92.6	1.4	11.16	1.50
Total	**248.9**	**4.0**	**25.84**	**3.48**
Total India	**6258.9**	**(100)**	**742.33**	**(100)**

It will be observed that 96 percent of the available freshwater fish was marketed in the British Indian provinces and only 4 percent in the Indian States.

Boats in the British Periods

The river crafts of the Gnges consists of the rafts and dug-out Canoes and Plank-built boats. The raft of inflated skins is much in use on the Ganges at Hardwar and on the tributaries and streams. The other forms of Gangetic rafts are found only in the marshy districts of the lowlands or are in use for fishing in tanks, jheels and other quiet waters. The commonest are those called bhela, bhera and bhewra. The former (Kalar bhela) is probably the most primitive form of raft evolved by prehistoric man.

The chatty or earthen pot raft, while almost equally primitive is still more ingenious. As used during the rainy season in Patna, Gaya and Hazaribagh districts, this float called gharnao or gharanai, is constructed usually of nine inverted earthen pots (gharas) arranged in rows of three. Single pots are also sometimes utilized to get across water channels while fishing with nets. The tigari or gamlaisin common use in Bengal. It is used by men, women and children for crossing streams when the country is flooded. The usual form of tigari is a basin-shaped hemispherical vessel made of burnt earthenware, about two feet and a half in transverse diameter. It is very like a small circular coracle is shape except that the bottom is not flattened. A short paddle with a long oar-shaped blade is used for paddling.

Dug-out canoes, called donga in Bengal, and ekhta in Bihar, are seen everywhere on the shallow tributaries of the Ganges. A large donga can accommodate three people sitting tightly packed in the butt end. The ordinary size is 11 to 13 feet in length, with a diameter at the fore-end of 2′ 8″ to 2′ 6″, while the narrow posterior portion is usually termed ekhta, a name evidently given with reference to this kind of boat being formed out of single log or tree trunk. It is also termed Sahia or Chhataki. In the Lohardaga district ekhta dug-outs are the only boats than can be used. The ekhta or salti runs from 15 to 30 feet in length, with a breadth of two fet or rather more, and about the same in depth. It draws 8 or 9 inches of water when laden. A single paddle is used, about 6 feet in length with a long parallel sides blade about 6 inches wide.

The plank built boats may be categorized into five categories:

1. Small passenger and fishing boats–usually called dinghies.
2. Rowing and racing skiffs.
3. Large fishing boats.
4. Travelling houseboats and ceremonial barges.
5. Cargo carriers.

Dinghis

The dinghies whether employed for fishing or for carrying passengers or goods with the frequent need to row against the current are characteristically low forward as it is the invariable custom for the rowers to be located at the fore end of the boat and not in the waist as the usual custom in Europe Small fishing dinghies do not usually indulge in the luxury of a cabin except in the rainy season. In the larger ones when present, it is always above the mast.

Fishing dinghies although differ considerably in detail and proportion and have their special local names not because of differences in the form but merely because of the kind of net used from them. The most common is known as Jalia dinghies. In form and construction all varieties of the Jalia dinghies are admirable and well suited to contend with the strong current and tides of the Ganges and its branches within tidal influence. The usual woods employed in the construction are sal, teak and jarul.

Rowing and Racing Skiffs

A dinghy is the long narrow boat propelled by numerous rowers or paddlers. One of the smallest is the bhaulia (bhauleah) propelled generally by four or six oars and with the small plank cabin. Another somewhat larger, is the fish carrying chip.

Large Fishing Boats

Of these the chhandi is a good example. The chhandi is used in draft net fishing in the lower and wider reaches of deltaic rivers. Another type of large fishing boat is the bachari principally used when fishing is done with the big caste and dip nets. It is longer but narrower than the chandi boat usually being 75 to 80 feet in length, with a beam 6 to 8 feet. A very remarkable boat is the mechno bachari used for the transport of live fish from the bhel districts of Eastern Bengal to Calcutta. The greater part of the hold is partitioned to form a live well, in which great number of fishes are transported in water.

Travelling Houseboats and Ceremonial Barges

In a deltaic region where waterways largely take the place of roads as the means of intercommunication, the traveling house boats or badjra are used to furnish shelter and accommodation to those using them. The badjra is the largest and most spacious of its kind, 30 to 50 feet in length, 8 to 16 feet in breadth and 4 to 6 feet deep. Except for some 8 to12 feet at the forward end, the deck space is entirely occupied by substantial flat roofed cabins. The rowers are located forward, two to four on each side. The oars (pathandnar) have long wooden blades tied to the ends of bamboo poles.

The ceremonial barges used by the territorial river lords are usually glorified badjras, gaily covered with paint and the mounted with a figure head of horse or bird. The high towering stern ends in some other fanciful figure.

Cargo Carriers

The craft used for heavy traffic on the Ganges are still full of human interest in comparison to the smaller ones that are used for fishing and ferrying purposes. To transport cargo in use are of numerous varieties which differ in detail according to the trade they specialize in and the local peculiarities of the particular waterways they usually traverse. The most common n the Hoogly are the (1) Bhar (2) Kola (3) Parulia bistupurai, (4) Bhadra kullai and (5) Khanjai Katti

Besides these, the balam carries cargoes between the estuarine ports. It is built after the same fashion as the large fishing boats of the Ratnagiri district in Bombay. The clinker-built ulkas, that have from time immemorial brought building stone from the Chunar quarries to Benaras and supply the city and the burning ghats with firewood, are almost the only river craft in the northern India that retain the extremely ancient custom of decorating the bows with 'eyes'. Each eye is fashioned in brass. The parts are nailed high up on each bow immediately behind the stern post. At the time of new moon (amavasi) as is the general custom on the Ganges a garland is hung from the stern -head. The worship is performed at the owe; a ghee lamp is lighted, camphor is burnt, the sacred conch is blown, and coloured paste is smeared on the stem head. Flowers, rice and milk are offered to the God whose protection of the boat and her crew is implored to keep them safe from the dangers of the river which are by no means negligible for a heavily laden boat when the river is in spate.

The various classes of boats, both great and small, playing on the Ganges and its tributaries, without any notable exception, form one compact body possessing certain fundamental features in common. We have the same multitude of high sterned, low prowed boats, steered by paddles. In the lager craft the same design is carried out on a perportionately greater scale, the high carved stern being retained while the forward parties kept low to permit the use of sweeps.

Chapter 5
"FISHERIES IN INDEPENDENT INDIA"

The research activities of fisheries in the country were modest prior to 1947. It was mostly species oriented and centered around the studies on the taxonomy of fish and descriptive natural history carried out by individual scientist working in isolated places. The union Ministry of Agriculture established central fisheries research station in the year 1947. The institute maintains a close liaison with the state fisheries departments and other institutions doing similar work.

The inland fisheries of India comprise two types of water; Freshwater and Brackish water. The fisheries of both these types of waters can be categorized into two types: (1) captures and (2) culture. The term capture and culture are some what self explanatory. In capture fisheries we harvest the fish crop without having to sow. The fisheries of rivers, estuaries, lakes and reservoirs fall under the domain of capture fisheries. In the case of culture fisheries one has to "sow the seed", nurse it, rear it and grow it to table size before harvesting. The principles of management of these two types of fisheries are quite different from each other.

Capture Fisheries

The capture fishery resources consist of 1,73,287 Km rivers, canals and irrigation channels. There are 13 lakh hectare beels, oxbox lakes and swamps. There are 29 lakh hectare reservoirs and lakes. India has 8129 kms long sea shore-line (GOI, 2003). Detailed investigations have been carried out on the fish and fisheries of Ganga, Krishna, Godavari, Narmada and Tapti river system and the Hooghly–Matlahan Mahanadi estuarine system by the Central Inland Fisheries Research Institute, Barrackpore. The biology of a number of species of carp and catfishes inhabiting these rivers and of estuarine fishes such as hilsa, mullets and prawns have been studied in detail. Studies on the hydro-biological factors affecting the fisheries of these rivers and estuaries have also been made. A rapid industrialization has posed the problem of pollution in Indian rivers and estuaries on account of discharge of industrial effluents and domestic sewage, making them uninhabitable to fish and fish food organisms. Detailed investigations in respect of extent and nature of effluents of paper and pulp, textiles, tannery, sugar and distillery, coal washery, cycle rim and vanaspati factories have been made in the Hooghly, Daha, Sone and Vhadra rivers. While these effluents have profoundly adverse effects on the fisheries of Daha, Sone and Bhadra, their effect on larger rivers with discharge is limited. Bio-assay experiments have indicated that while the paper and pulp effluents are non-toxic to fish and fish food organisms; tar, distillery, tannery and cycle- rim factors wastes are highly toxic. Studies of the effect of sewage on the fisheries of Kulti estuary have indicated that a large part of the estuary has become uninhabitable for fish and fish food organisms.

Fish and fishery surveys in connection with the construction of river valley projects such as Tungabhadra, Hirakund, D.V.C. Rihand, Nagarjunsagar were undertaken with a view to formulating suitable measures for the development of fisheries in the reservoirs and river basins. Surveys have been conducted to locate breeding grounds of carps. The centers for the collection of fish seed in the riverine stretches of many reservoirs such as Pilwa and Kotwal (M.P.), Tilaily and Panchet Hill reservoirs of the D.V.C. and Budha in Bihar have been established.

Experimental fishing has been undertaken recently in many reservoirs in different regions to develop deep-water fishing. Work on experimental fishing undertaken by FAO experts has resulted in the improvement of the local nets, designing of new nets, use of echo sounder for charting the bottom and locating fish shoals, and new methods of fishing as light and electrical fishing.

Fisheries of Freshwater Rivers and Canals

The river system can be classified in two main groups (1) the Himalayan rivers and the extra–peninsular rivers which comprise (i) the Ganga system (ii) the Brahmaputra system and (iii) the Indus system, and (2) the Peninsular rivers which comprise (i) the Eastern Coast system, (mainly composed of the Mahandi, the Godavari, the Krishana and the Tapti). The snow and rain-fed Himalayan rivers are characterized by a complicated flood regime, heavy silt load, marked seasonal variability in volume, shifting of their course and heavy bank erosion. In contrast, the Peninsular rivers, only rainfed, are torrential with a well defined stable course, lesser silt load and rocky terrain except in the deltas. Commercial fisheries in the upland water of the Gangas system are virtually non-existent due to exploitation problems and poor communication link. The snow trouts (*Schizothorax* spp. and *Schizothoraichthys* spp.) and certain catfishes are common in the headwaters of the Ganga below 1067 m Mahseers (Tor tor and T. Putitora), certain cyprinids (*Labeo dyocheilus* and *L. dero)* and the catfish (*Bagraius bagarius*) form the food fish. They descent to warmer waters of the plains and contribute significantly to the commercial fisheries. The stretch of the Ganga (1600 Km) between Hardwar and Lalgolaghat provides the richest source of freshwater capture fisheries in India. The commercial fishery is composed of major carps (*Cirrhinus mrigala, Catla catla, Labeo rohita* and *L. calbasu*) and catfishes (*Wallago atttu, Mystus aor, M. seenghala, Silonia silonia, Pangassius* and *B. bagarrius*). The anadromous Hilsa (*Hilsa illisha*) contribute significantly (30–60%) to the total catch during season of its migratory run. This system provides more than 90% of total carp spawn production in the country.

The headwaters of Brahmaputra harbour Mahseers and *B. bagarius*. A total of 126 species from 26 families have been recorded of which 41 are of varying commercial importance. The commercial catch in the middle and lower reaches is dominated by larger catfishes, carps (*Labeo gonius, L. rohita*), hilsa and freshwater prawn.

On the Indus system, the headwaters of rivers Beas and Sutlej harbour mahseers, snow trouts, cyprinids (*Labeo dero, Garra gotyla* and *Barillius* spp.) and the exotic brown and rainbow trouts. The Beas and the Sutlej contain indigenous carps and catfishes are akin to the Ganga drainage.

In the headwaters of Mahandi, Mahseer (*T. mahanadicus*), *B. bagarius* and *Labeo dychocheilus* are abundant. In its middle an lower reaches, Gangetic fauna occurs in addition to a few penisular forms (*Labeo fimbriatus*). The upper reaches of the Godawari harbour game fishes but do not support rich fishery. Carps (*L. fimbriatus* and Gangetic major carps). Larger catfishes (*Mystus* spp. *W. attu*, *Silonia childreni* and *B. bagarus*). Hilsa and macrobrachium constitute the commercial fishery. The headwaters of the Krishna and its tributaries, interspread with shallow rocky shelters, abound in rich fish fauna as compared to the wide sandy middle stretch. The Cauvery exhibits substantial variations in fish fauna. The game fishes (*Tor khudrae* and *T. mussullah*) occur all along its length except in the deltaic region. The Cauvery fish fauna differs significantly from that of the Godavari and the Krishna though hilsa occurs in the lower reaches. The commercial fishery is constituted by mahseers, *Puntius carnaticus, P. dulcius, L. kontius, Glyptothorax madraspatanum, Mystus* spp. *P. pangasius, W. attu* and *S. silondia* together with the Gangetic major carps transplanted from the Ganga System.

The commercial catches in Narmada consist of carps, (*L. fimbriatus, L. calbasu, C. mrigala*), catfishes (*Rita pavimentata, Mystus* spp. and *W. attu*), Murrels and featherbackes; whereas *T. tor, L. fimbriatus, L. boggut, Mystus* spp. and *W. attu* provide for the commercial fishery of river Tapti.

Brackish Water Fisheries

The area of these in India is estimated at 12.35 lakh hectares comprising 293 km stretch of the Hoogly proper upstream from Sandheds and the other estuaries of the region, Matlah, Saptamukhi and Haldia. This estimated area also includes coastal waters and marginal saline waste lands exposed to tidal inundation. The fisheries in the system are typically of a multi species type, exploited principally by very small meshed bag nets. The fisheries produce largely smaller forms, juveniles and young ones. The ecology of the system has undergone major changes after commissioning of the Farakka Barraage. The increased freshwater flow has reduced turbidity, lowered salinity, pushed the gradient zone seawards, increased flow velocity, raised abundanc of plankton and has improved the general habitat, especially with regard to the problems of pollution caused by industrial waste effluents discharged into the Hooghly. Some of the typically euryhaline species whose catches had been showing a decline in spite of overall increase in landings, appear to have significantly picked up. Amongst these mention may be made of *Pama pama*.

The hilsa fisheries of the system has been fluctuated heavily. Although, above Farraka Barrage in Ganga river, the fishery has collapsed, in Hoogly, it is yielding at a higher level of about 2,000 t/yr. Apart from above, the system yields a sizable quantity of seed of marine prawn P. monodon, Bhetki and Mullets for brackish water aquaculture.

Mahanadi System

The mahanadi estuarine complex covering an estimated 30,000 ha. area; the brackish water of Hukitola lake of 10,000 ha and Jatadharmohan of about 2000 ha. has been estimated to produce a marketable surplus of about 550 t/yr. The principal fisheries of this estuarine complex, after the decline of Hilsa in 1961–1962, has been the mullets, dominated by *M. cephalus*, accounting for 37–47% of total landings.

Estuaries of North West Coast

Narbada and Tapti form the two main estuaries in west coast. The fisheries of these estuaries have not been well estimated. Hilsa fisheries in post- monsoon months are known to be sizeable *i.e.* 767 t/yr.

Fisheries of Freshwater Reservoirs and Lakes (Lacustrine Fisheries)

The reservoirs may be broadly divided into three classes:

1. Stocked with major carps, which have established themselves into self regenerating populations.
2. Stocked with major carps, which are yet to establish themselves into self regenerating populations.
3. Not stocked or stocked with major carps and are not capable of generating into auto-stocking population.

Studies conducted in DVC reservoir have indicated that some stocked species have established themselves, while others have not done so. Against estimated average yield rate for 6.5 kg/ha./yr for Tilayia, 1.9 kg./ha/yr for Maithon, 6.4 kg/ha/yr for Konar and 2.7 kg/ha/yr for Panchet during 1962-68, the productive potentials are placed at about 100 kg/ha/yr at present. Studies indicate the production potentials of Bhavanisagar, Rihand, Gobindsagar, Nagarjunsagar and Ukai reservoirs as 118, 40, 276, 88 and 220 kg/ha/yr respectively. For Stanlway, Bhavanisagar, Sathanur, Krishnagiri and Amarawvathy reservoirs of Tamil Nadu, the yield rates corresponded to 62.9, 27.8, 77.0, 20.5 and 160 kg/ha/yr respectively in 1968.

Fisheries of Brackish Water Lagoons

The three principal brackishwater lagoons are the Chilka Lake (116,000 ha) Pulicat lake (77,700 ha) and Kerala backwaters (50,000 ha.).

In Chilka lake, the yield is reported to be declining from 4,700 ton in 1964–1955 to 3,000 ton in 1964–1965 and 2,740 ton in 1970–1971. Earlier, the yield was about 3,600 t. during 1948-1952. The principal fisheries are of *M. cephalus* (13%), *L. macrolepis* (3%) lates calacarifer (6%), *Mystis gulio* (8%), *Polynemus terradactylus* (6%), *Hilsa ilisha* (4%) amongst fishes and *P. mondon* (6%) and *P. indicus* (18%) amongst prawns.

In Pulicat lake, it was estimated to produce–1173 t per year between 1965 and 1970 as against reportedly 2,678 t in 1945–1946. *P. indicus* and *M. cephalus* contributed the principal fishery.

Kerala backwaters

Reports put the yield around 14,000 to 17,000 t/yr in addition to 88,000 t clams and 170,000 t of molluse shells. Prawns constitute 60–70% and mullets 11% of the catch at Vembanad Lake. In recent years the prawn fishery has shown a marked decline which is attributed to over fishing in the backwaters as well as in the sea.

Cold Water Fisheries of Hills Streams

The fisheries resources in the cold water streams of the Himalayas in the north and Nilgiri Hills in the south are not well estimated. The total length of the streams excluding eastern India is estimated at 3,711 km.

High altitude cold water reservoirs sustain fisheries of some importance. They vary in size from 1.8 to 93 ha in area, except Wular (6,400 ha) and Dal (1,490 ha) in Kashmir. There are two brackishwater cold were lakes too, Pongar Tso (290 ha) and Tso Moriri (2120 ha) in Ladakh aea. The fisheries are poorly developed in these water loader.

Culture Fisheries

The culture fishery resources include 42 lakh hectare of water including 22.54 lakh hectare of freshwater tanks and ponds.

Breeding of Carps

As the Indian major carps do not ordinarily breed in confined waters, their young ones have to be collected from rivers and streams with uncertainty of availability and purity. Although investigations carried out with a view to study the factors responsible for their spawning have not yield conclusive results, great success has been achieved in breeding them in artificial conditions. As the breeding of carp in rivers or bundhs is subject to several climatic and environmental factors which are so far beyond human control, work on inducing the fish to breed in confined waters of ponds and even in cisterns was taken up by the Central Inland Fisheries Research Institute. The Indian major carps, many medium sized carps and other food fishes have been induced to breed by the administration of pituitary extract. As a result, a large number of fish breeding centers have been opened in the country where carp breeding is undertaken every year during the monsoon and a large quantity of fish-seed is produced. Two Chinese carps, *i.e.* grass carp and silver carp have also been bred by the administration of homoplastic or heteroplastic pituitary extract injections.

Success in induced breeding of fishes has opened up a new vistas of research through selective breeding and hybridization making it possible to raise hybrids of superior qualities then either of the parent species. Several inter–specific and intergeneric hybrids of Indian carps have been produced at the central inland Fisheries Research Institute.

An outstanding success has been achieved in the hybridization of the Indian carps.

Inter Specific Hybrids

Male Parent	*Female Parent*	*Hybrid*
Labeo rohita	*Labeo calbasu*	*Rohu-calbasu*
Labeo calbasu	*Labeo rohita*	*Calbasu-rohu*
Labeo bata	*Labeo rohita*	*Bata-rohu*
Labeo bata	*Labeo calbasu*	*Bata-calbasu*
Labeo-calbasu	*Labeo gonius*	*Calbasu-gonius*

Investigation on the influence of light and temperature on the reproductive cycle of carps to bring about maturity at any convenient time to the year are in progress at the Central Inland Fisheries Research Institute. Similarly, experiments on the effects of feeding vitamin E (Ephynal) to bring about an early maturity in fishes are also in progress.

Exotic Fish Culture

The need for pond–breeding and fast growing species was long felt in our country. Te first exotic species to be introduced in recent years was the controversial Tilapia massambica, a prolific pond breeding cichlid which can survive extremely variable environments. Tilapia was found to affect the major carps adversely and hence was objected to as a compatible species for culture in carp ponds.

A new strain of common carp, *Cyprinus carpio* var *communis* was introduced from Bangkok and simple technique was developed to bred the fish in cloth hapas on a commercial scales. The Prussian strain, *C. carpio* var *specularis* introduced in the Nilgiris in the late thirties and which hitherto bred only in the hilly regions was also successfully made to breed in the plains following the same technique as for the Bangkok strain. The former has been found to grow faster than mrigal and rohu and almost as fast as catla under Indian conditions. Two more exotic fishes, grass and silver carps, introduced into India in 1959 and have been stimulated to breed by pituitary extract injections. Their progeny attained maturity in the very first year as against three years in their native habitat.

Brackish Water Fish Culture

Vast potentialities for the development of brackiswater fish culture exist in India, but the lack of data on scientific methods for farming inhibited progress in this direction. In recent years, studies of the hydrobiological and soil conditions of different types of brackishwater fisheries, growth requirements of blue green algae, effects of salinity on nitrogen transformation and the type of fertilizer for brackishwater soils, have been made. Methods of culturing brackishwater fishes and prawns in the paddy fields of West Bengal and Kerala respectively have been examined and suggestions for increasing fish and prawn production were given.

Exploratory surveys for the location of collection centers for the seed of some important cultivable brakishwater fishes like Bhetki, Mullets, and Prawns have helped in locating various centers in the Hoogly estuary. Indigenous methods of collections and transport of Mullet seed have been studied and suitable suggestions for the their improvement was made. Methods of distinguishing the fry of different species have been described to enable selective stocking of brackishwater forms for ensuring increase production. Studies of anesthetizing the mullets for transport without mortality and the food preference of its fry and fingerlings are in progress. An experimental fish farm at Kakdwip (West Bengal) is under construction where large scale studies on the various requirements of scientific fish farming of brackiswater fishes and prawns would be undertaken in an organized way by the Central Inland Fisheries Research Institute.

Cold Water Fish Culture

The breeding grounds and the spawning season are the important factors for indigenous species of coldwater fishes–*Tor putitora* and *Oreius plangiostomus* have been determined in Himachal Pradesh. An ecological survey of the Mahseer and trout living in the streams of Himachal Pradesh and Kashmir has been completed. Experimental hatching operations undertaken recently at the Trout Farms in Himachal Pradesh and Kashmir have revealed the high rate of mortality during the various stages of trout culture.

Development of the Inland Fisheries Resources

The Inland Culture fisheries resources are classified as:

1. Freshwater ponds, tanks and swamps.
2. Brackish water ponds and impoundments.

Freshwater Ponds, Tanks and Swamp Resources

These resources abound in the plains and Assam, Bihar, West Bengal, Orissa and coastal Andhara Pradesh. Highly divergent estimates of this resources class have appeared from time to time. The fish seed Committee of Government of India reported the ponds and tanks cover 2.54 million hectares are suitable for carp seed stocking. The National Committee of Agricultural made no distinction between ponds and tanks. Pilot studies conducted by National Samples Survey organization, Central Inland Fisheries Research Institute and others have shown that about 85–90% ponds in West Bengal are utilized for one or another form of fish culture and exploitation.

The yield rate is dependent on culture practices. The NSSO survey in Murshidabad and Thanjavur district have shown yield rates in the range of 100–700 kg/ha/yr. Surveys conducted by CIFRI show the yield rates from 100 to 2,000 kg/ha/yr under various conventional culture practices in West Bengal. A national average of 600 kg/ha/yr under conventional culture has been obtained.

Brackish Water Ponds and Impoundments

A preliminary survey (1961–1962) of existing fisheries in Sunderbans are area of West Bengal, showed about 150 bheries (Bhasa-badha), covering a total of 9,572 ha and yielding 3,000 t, mostly made up of mullets, bhetki and prawns. The average yield rate worked out to be 313 kg/ha/yr. Some of the latest data show the total available are as 1,712 million hectares. The utilized area at present for brackishwater aquaculture is estimated at 88 ha in Gujarat, 4,800 ha in Karnataka, 5,117 ha in Kerala and 20,000 ha in West Bengal. This shows only 1.8% utilization of the available area.

Transplantation of Fishes in Independent India

To transplantation of fishes in India has been mainly for food, sport, mosquito control and ornamental purpose. The following food fishes have been transplanted for purpose of experimental culture in post independent India.

1. Mirror carp (*Cyprinus carpio* var *specularis*)
2. Scale carp (*Cyprinus carpio* var *communis*)
3. Silver carp (*Hypophythalmichthys molitrix*)

4. Grass carp (*Ctenopharyngodon idella*)
5. Tidapia (*Tilapia mossambica*)
6. *Puntius goninotus* (the tawes)

Mirror Carp (*Cyprinus carpio* var *specularis*)

In 1947, this species was transplanted for the first time in Kumaon hills (U.P.) from Nilgiris and by the end of 1954, it was introduced in Nainital Lake and other lakes of Kumaon Himalayas. The fish was introduced on April 18, 1955 in "Pucca Tank" at Nattan (Sirmor district, Himachal Pradesh). From Himachal Pradesh, mirror carp has been transplanted to Punjab, Jammu and Kashmir, Madhya Pradesh, Rajasthan, Delhi, Bihar, Manipur and also Sikkim (Jhingran, 1975). In 1949 a small consignment of fingerlings of mirror carp was sent to Bombay from Ootacamund and introduced into a lake at Lonavala.

Scale Carp (*C. carpio* var *communis*)

It was introduced from Bangkok in 1957 at the Central Inland Fisheries Research Farm at Cuttack. The species is a prolific breeder and has been successfully established throughout the country.

***Hypophtalmichthys molitrix* (The Silver Carp)**

The fish is a native of China and Amur Basin (U.S.S.R.). It was introduced in India from Japan in 1959 at the Cuttack Rersearch Station of the Central Inland Fisheries Institute (ICAR) and has been successfully induced bred and propagated to all parts of the country. The fish has recorded growth of 2 kg in 10 months and 4-5 kg in 16 months in Kulgarhi reservoir.

***Ctenopharyngodon idella* (the grass carp)**

It is also a native of China and Amur basin (U.S.S.R.) and is widely cultured in many parts of the world. This is one of the most efficient biological agents for aquatic weed control. It grows very fast and has the 'shortest food chain'. In India it was first introduced at the Cuttack Research Station along with silver carp in 1959 from Hongkok.

***Tilapia mossambica* (The Tilapia)**

It is a native of Africa. It was first introduced in India in 1952 by the Central Marine Fisheries Research Institute in Madras. This is a prolific pond breeder, grows fast in fresh and brackish waters and thrives well at high altitude (1,000 m) but cannot withstand a temperature below 10ºC. The species has been reported to grow upto 850 gms in freshwater and 450 gms in brackish waters. In intensive culture experiments at Cuttack, production of about 5,000 kg/ha/yr and in sewage fed ponds in West Bengal above 5,000 kg/ha/yr was achieved.

Puntius gonionotus (The Tawes)

This is pond breeding fish which exercises some control on aquatic weeds. It was introduced in India at the Kalyani farm in West Bengal by the Central Inland Research Institute (ICAR) in 1972 from Indonesia. It has been bred at the farm and is being cultured for experimental purpose. It is a slow growing fish in comparison to major carps.

System of Freshwater Fish Farming

Composite Fish Culture

Extensive researches on the compatibility of exotic species of carps in mixed culture with Indian major carps were carried out at the pond culture division of the Central Inland Fisheries Research Institute (CIFRI), Cuttack. The fish production increased much more when the pond was stocked with 6 species combination *viz.*, Catla, rohu, mrigal, silver carp, grass carp and common carp compared to the production obtained either with catla, rohu mrigal combination. The high yielding combination of 6 species of carps is termed as composite fish culture. The combination has spread all over the country through a coordinated research project of the ICAR.

Sewage Fed Fish Farming

Detailed investigations towards the scientific utilization of sewage in increasing fish production are being carried out by the CIFRI with promising results. Polyculture of Indian major carps in oxidation ponds is also being carried out at Bhopal and Nagpur with encouraging results (Sinha and Ramachandran, 1985).

Air-breathing Fish Farming

The common air breathing fishes of commercial value are :

Murrels	Giant murrel	*Channa marulius*
	Striped murrel	*Channa striatus*
	Spotted murrel	*Channa punctatus*
Magur	Walking fish	*Clarias batrachus*
Singhi		*Heteropneustes fossils*
Koi	Climbing perch	*Anabas testudineus*

Efforts made under all Indian Co-oriented Research Project to develop suitable technique for air breathing fish culture have given encouraging result (Sinha and Ramchandaran, 1985).

Paddy- cum Fish Farming

In West Bengal and Kerala there is a common practice of the utilization of paddy fields for culture of brackish water prawns and fishes. The culture of *Channa striatus* in paddy field for 122 days yielded about 113 kg of fish per hectare and at the same time increased the yield of paddy by 7 to 13 percent (Iyengar, 1953).

Encouraging results were obtained at Aquaculture center at Dhaulai on rearing of common carp fry along with China 1039" paddy variety (a hybrid). The rearing period lasted for 99 days. The average size and weight increments of the fish were 36 mm/12.4 gms with 88 percent recovery. Comparable results were also obtained in a paddy plot of 0.16 ha. at the Central Rice Research Institute, Cuttack.

Fish Farming in Integrated System

An integrated farming of fish culture, dairy and poultry has been traditionally practiced in different parts of the world with varying degree of success. In India the system of fresh water fish culture has assumed greater significance presently in view of its potential role in recycling of organic wastes and integrated rural development. Besides the cattle farm wastes, which have been traditionally used as manure for fish pond, considerable quantities of wastes from poultry, duckery, piggery and sheep farming are available. The latter are much richer in nutrients than cattle wastes and hence smaller quantities would go a long way to increase fish production. The work of the operational Research Unit on composite fish farming at Krishnanagar, West Bengal showed very promising results in fish production with pig manure.

Inland Fishing Gear

The Central Institute of Fisheries Technology was established in 1957. Since then the institute has controlled a lot to the improvement of the inland fishing gears.

The fishing gear of the river systems shows certain divisions along its entire course. (Chacko (1955) divided it into three sections namely, upper, middle and lower. In the upper reaches, where normally a swift current prevails, cast nets are used. In the middle reaches seine nets and gil nets are equally important. Joseph *et al.* (1965) considered seine nets to rank first in the middle reaches of the Brahmaputra river system. Cast nets, stick held seine nets lefe–nets barriers are of secondary importance.

Fishing operations conducted at Gobindsagar reservoir (Punjab) have resulted in locating a number of productive fishing grounds in the upper and lower reaches and evolution of the design of a suitable simple gill net. The design of this net has been adopted by state fish units and commercial fisherman. Similar investigations in the Hirakund reservoir (Orissa) have shown potential fishing ponds in the middle reaches of the reservoir and the relative superiority of the gill nets in having a greater entangling capacity. In the Sunderbans in West Bengal collaborating with CIFRI and the state Department of Fisheries, suitable gears have been designed and recommended. Similarly, surveys of Mathura, Kalyani and Kulia Bheels have enabled the state Fisheries Department to recommend the use of suitable fishing implements.

The fishing investigations conducted at the Gandhisagar reservoir (M.P.) have obtained the institute to evolve the design of a gill net for the effective exploitation of *Labeo calbasu* and endemic population of the reservoir. A preliminary survey of existing fishing gear and the fish population of the Malampuzha reservoir (Kerela state) was undertaken, and a detailed scheme of investigations drawn up and furnished to the state Fisheries Department.

Studies of the fishing gear and methods provided the essential background knowledge for understanding improvement and exploitation of the fishery. George (1971) attempted to present the technical characteristics of the fishing gear and methods in vogue in different Inland water systems in the country. Over sixty types of fishing gear, belonging to ten classes, were observed and described by him. He also described the design, construction and method of operation in detail. He followed the classification of Von-Brandt (1958, 1964) and the designs were presented according to the pattern in the F.A.O. catalogue of fishing gear (1965).

Name of Centre	*Systems of Water Body*	*Name of Station*	*Gear Surveyed*
Calcutta	Hoogly Estuary	Diamond Harbour	Chondana jal Kajolgouri jal, Illishmach jal and Bekti jal
		Barrackpore	Beem jal and Kuto jal
		Medagachi	Borsi, Vashakona Schiki jal, Bua jal, Mahal jal, Burl jal, Kepla jal, Dholy jal, Chhandi jal and Shangla jal.
Allahabad	River Ganga and Yamuna	Sadiapur	Jali, Do-Dandi, Maha jal, Karia jal, Derwari, Chhanta jal, Gopal jal, Kuriyar, Thiar and Kamel
		Jhusi	Gochail jal
		Lawain	Bandel
Udaipur	Jaisamand Lake	Jaisamand Lake	Gasita jal, Tapar, Dhaiya, Bariya, Kathariya and Phasula
Hoshangabad	River Narmada	Beelpur	Daman, Kumni, Pelni, Feka, Kadela and Thangadi
Vijaywada	River Krishna	Vijaywada	Kondala, Polsa Vala and Polsa Vala
Hospet	Thungabhadara Dam	Thungabhadara Dam	Rangoon net and Udu valai
Banglore	Bellendur Tank	Ballendur Tank	Singidi valai, Glippu valai, Gangoon net and Udu vallai
Mettur	Mettur Dam	Pannavadi	Rangoon net (Kenda valai and Thoppa valai) and Udu valai

Line Fishing

Long lines with small hooks are operated for predatory fishes in estuaries, rivers and reservoirs. Borsi of Medagochi is a bottom set line, while Daman of Hoshangabad is a surface set one. The length of the main line ranges upto 1000 meters with 450–500 hooks, which are in turn fixed to the main line at regular intervals by snoods.

Fish Traps

Fish Barrier

In principle this consists of leading the fish by means of bamboo screens and their final capture by lever nets. The method is practiced along the river Ganges in the vicinity of Allahabad. The gear consists for two parts namely, the bamboo barriers and the Bandel.

When operated for ascending Hilsa the barrier consists of three sections, the main section being set in an inclined direction across the river making an angle of 30–40° to the shores of the river: the other two sections are set in the form of "V" with the apex of the "V" facing the ascending hilsa.

The barrier in case of descending hilsa is set in the form of 'V'; but the arms of the 'V' are comparatively longer than the former. Ganch is fixed between the arms of the 'V'. The shape of the net is triangular with the mouth supported by frames, each 6.6 metes in length. The webbing is rectangular and loosely mounted on the frame.

The screens are fixed prior to the commencement of the season while the lift net is fixed only after the season starts. The migrating hilsa, guide by the screen, enters the lift net striking on the webbing of the net and causing it to vibrate. The man on watch immediately lifts the net collecting the fish by small hand nets. The season for operation of this gear extends from January to June.

Buoyed Traps

The principal is similar to the trap described earlier, the main difference in the design is that the bamboo screens are replaced by a net wall and the fishes are captured in the blind pockets of the trap. The gear Vasha Kona comprises of the leader and a chamber or trap proper. The leader is rectangular, 30 to 35 meters in length and 7 to 9 meters in depth. The chamber has the shape of one half of a conico-cylinder in the longitudinal section, with the distal end narrowing and approaching the shape of a 'V' which finally terminates in a blind bag.

The trap has two pairs of flappers, one pair following the other. At the proximal ends of flappers are two bamboo poles which keep apart the walls of the chamber, the front and lower margins of the flapper are laced to the walls of the chamber. Connecting the distal end of the first pair of flappers there are 17 pairs of thin cotton ropes. The head rope of the first pair of flappers continues beyond its limit and extends back to the second pair of flappers, then it is joined into the ropes of the chamber. Beyond the second pair of flappers is the cod end. The chamber is followed by another rectangular section of net identical to the ladderdescribed earlier. In larger nets, there are two chambers and three ladders.

The net is played form the aft of a boat manned by a crew of 4 to 5 persons. The net is laid across the river in an inclined direction with the mount of the trap facing the direction of migrating hilsa. The net usually extends the entire breadth. The fish guided by the leader enter the trap and are finally caught earlier in the cod end or occasionally between the flapper and wall of the chamber.

Fish Pot

This gear is popular in the upper reaches of rives and rivulets where flow of water exists. Kumni of Hoshangabad has a cylindrical shape with one end tapering to a blind end where the tapped fishes accumulate. The trap is 60 cms in length and 75 cms in circumstances. The entire trap, including the body and the value, is constructed out of bamboo strips and fastened by cotton twine. The value is trapezoidal in shape. It is fixed to the mouth of the trap.

The trap are set across the current, a number of them in rows with mud or pieces of rocks in between to secure them to the ground. They are set in the evenings and lifted in the mornings. The catches are varieties of small fish.

Aerial Traps

Rafts or net walls are erected for inducing the fish to make long leaps and in this process they are captured.

Raft Trap

Raft traps are operated for mullets, which are known to take long leaps at the sight of an obstacle. The rafts are constructed by securing plantation trunks together. Twings are kept on the surface of the trap, they are either moored or drifted; in the former case fishes are driven towards the raft while the latter the fishes that come in the way of the raft are caught. Twings spread on the raft prevent further leaping by the fish. The method of practiced in the rivers Ganges and Yamuna near Allahabad.

Varandah Net

The gear is found in the 'bleels' of the West Bengal. It consists of a net wall erected in an oblique manner facing the direction of movement of the fish. The fish at the sight of obstruction leaps and is caught in the folds of net wall. The net is formed by securing webbing to bamboo poles driven to the ground in an oblique manner. The poles form the corners of a polygon. This webbing consists of two portions; namely, an upper or above water portion and a lower or underwater portion. The upper selection of webbing has a fold towards its lower periphery. When it leaps and hits the upper section, the fish falls down into the fold. The catches are mainly major carps.

Bag Bets with Fixed Mouth

Push Net

The type is found in all water systems. A typical one consists of a 'V' shaped frame of approximate size of which the webbing is attached. The frame is constructed by securing bamboo poles or light wooden pieces at an angle of 45°, the two arms of the 'V' forming the sides of the mouth and the side opposite the angel being the base.

The webbing is divided into two parts namely, the belly and the cod end, both rectangular in shape. The cod end may be partly and completely fabricated of fine cloth, while the belly is fabricated with small rectangular pieces of webbing. A row of selvage may or may not be present between the webbing and rope. The rope is in turn tie to the frames at regular intervals. A progressive increase in mesh size is noticed from the cod end towards the belly.

Apart from the general design described above certain variations are also noticed. The presence of a separate pieced of webbing along the base of the mouth is noted in Schiki jal. This extra webbing is said to prevent the fishes from falling back into the water during hauling. Instead of 'V' shaped frame, rectangular frames or noticed in Pelni and Singidi valai. A rectangular pieced of webbing is inserted at the junction of cod end and belly to give the net the required shape as in Pelni and Jali. In Singidi Valai, covering of webbing fixed to the frame. This contrivance prevents the entry of weeds into the net.

Scaling methods are used for diving fish to the net when the net is operated as a stationary bag net. Scaring may be done by disturbing water with 'Y' shaped poles. The catch consists of small fishes and prawn. The net is operated during all the seasons, especially during morning and evening. The net is also operated as a lift net from boats as well as for collecting small fishes that shelter under the floating weeds in bheels.

Stow Net on Stakes

Spawn collecting nets belong to this group. The net has the shape of trawl. The Bua jal of Hoogly estuary is fixed by two parts of stakes. The net is secured to the stakes at the free end of the wings and at the junction of the wind to the bag. The length of the bag varies from 5 to 6 meters and that of the cod end upto 7 meters. The head and food ropes are 5 to 7 meters long. The material of the webbing is cloth. The stakes are driven in very shallow water, 1 to ½ meters in depth. The net is secured to the stake on both the head and foot rope. The net is operated only during flood season and remains in water through out the day. The eggs and spawn collected are periodically removed.

Slow Net on Anchor

The type is common in the Hoogly Estuary. The Beem jal consists of the bag tapering into cod end and two wings tapering towards the free end. The mouth of the net is kept open during operation by bamboo poles. The bamboo poles, two in number, are fixed between the head and foot ropes at the joining of the bag to the respective wings. Usually two floats, sealed kerosene oil tins, are used. There are secured either in the middle of the wing or at the free wend. At Diamond Harbour, large eastern pots are used as floats.

The net is operated from large boats across the estuary with the mouth of the net against the flow of the current. It is set prior to the tide and hauled in after the tide. Catches are mainly Scaienids, Gobids, Polynemids and prawn. The Thor jal of Diamond Harbour remains in water for one week. At Sunderbans, the net is operated between creeks. Behundi jal also belongs to this group. A big meshed net of this type, fabricated out of hemp, is operated at Diamond Harbour.

Do-dandi of Allahabad has two sticks at either extreme of the net while Kondala of Vijaywada has numerous 'cross bars'. In

Kondala the width of the webbing is greater compared to the width of the mouth opening, hence a small bag is formed in the net.

Beach Seine Without Peripheral Pockets and Bag

Considerable variations are noted in the general size and mesh size of the net determined by the area of operation and fishes caught. The larger nets are divisible into three parts, namely the two wings and the middle landling part or 'bag', the landing part is differentiated from the wings by a smaller messy. These features do not hold good for small beach seines like Iluppu Valai.

Haha jal and its various modification represent the common beach seines in vogue. Karia jal and Derwari of Allahabad are respectively big meshed and small meshed edition of Maha jal. The smallest mest size is noticed in Maha jal of Hoogly Estuary, where cloth is substituted for webbing. The wings, in usual cases, are equal; but in Derwari the wings are unequal, the shore wing being shorter.

All the beach seines are operated in an identical manner. In the final stage of hauling the foot rope is manipulated in such a way that it reaches the shore prior to the head rope, but never risen form the bottom. The operation of Iliuppu Valai differs form the above description as the net is dragged by two men only. Fishes are scared to the net by two men beating on the water surface.

Beach Seine with Peripheral Pockets but Without Bag

These nets differ from the previous type of beach seines in the presence of one or more row of pockets or folds towards the lower periphery of the net, and the fishes are collected in these folds or pockets. In Posals of Vijaywada the lower periphery of the webbing is only folded and fixed at regular intervals. In Ghasita jal of Udaipur and Chanta jal of Allahabad there are two rows of pocket, the pockets are formed by folding the lower periphery of the webbing along with the foot rope and sinkers.

In those nets where there are two rows of pockets, the two lower layers of webbing are folded to form pockets. Instead of securing the folded webbing at regular intervals as in Posala vala, the folded portion is laced to the main webbing at regular intervals in Ghasita jal and chhanta jal, thus sectioning the entire fold into a series of pockets all along the length. The direction of lacing in both the rows is parallel in Ghasita jal and zig-zag in Chhanta jal.

The mouths of the pockets are provided with sinkers. The operational details are same as that of other beach seine except the bead and foot ropes are not manipulated to concentrate the catch in the center.

Boat Seine

Buro jal of Hoogly Estuary is included under this group. The net has a bag and a pair of short wings. In construction it is similar to a bag net and operation is that of a typical boat seine. The warp of one side is fixed to a pole driven in shallow water and the net is payed from the boat in the form of an arc. On returning towards the starting point one more bamboo pole is driven and the boat is in turn fixed to these two poles. The net is then hauled up from the boat, dragging it at the bottom. Catches are mainly cat fishes, gobies and prawn.

Surrounding Gear

The principle here is to surround a detected shoal or a probable shall by a net or other contrivance. Chaundi jal of Allahabad is payed in a circular fashion in shallow water. The area of the circular is later reduced and the enclosed fishes are captured by a lantern new Kuriyar. The scare line is payed in the form of an arc covering the entire breadth of the river. Towards the final stages of operation the scare line is fixed around the three bamboo poles confining the fishes within the limit of the triangle. The fishes are then gaffed. The catch by these methods is mainly *Mystus* spp.

Drive-in Nets

Although these nets are similar to the stick–held seine net, in operation the fishes are actually driven into the net. In Gopal jal of Allahabad, the free end of the cross bars extends beyond the foot rope and this facilitates in fixing the net to the ground. During operation the net is fixed in the form of 'U' and the fishes are driven into the net by scare lines constructed by twisting old nets. The length of the scare line is gradually reduced. Towards the final stage of operation, the two ends of the net are brought together and the confined fishes are captured by lantern nets. The catches in this case are also Mystus species.

Falling Nets

Coverbots or Plunge Basket

These are gears which are cast on the fish and the fishes are taken form above. They are usually of wicker construction with the main opening at the top. Taper of Jaisamanbad lake has the shape of a bell. The material in bamboo strips secured by coir rope. Catches are mainly *Ophiocephalus* spp. Similar baskets are in use all over the country. Ooatha or Ottal is the type found in Kerala, Madras and Mysore.

Lantern Nets

While the fish enclosed in the cover pot has to be further removed and collected, in lantern net this problem is solve dby entangling or gilling them in the webbing. In Kuriyar of Allahabad, the wickerwork is substituted by webbing. The webbing, is held in form by six bamboo poles radiating from the apex. The webbing is in the shape of a cone and is in turn fixed along the lower margin to the free ends of the bamboo poles. The top end of the webbing terminates in a cotton rope, which is held fast during shooting and then let loose to eatable the enclosed fish. Larger varieties of this net with three radial poles and a basel ring are found in the bheels of West Bengal.

Cats Nets

While the use of coverpots and lantern are limited to a certain depth, the cast nets are operated relatively deep. Dhaiya bariya and Kathariya of Udaipur belong to this category.

Nets Without Closing String with Peripheral Pockets

In these nets the fishes are collected in the peripheral pockets. The pockets are formed by turning the weighted lower edge of the net and fixing it to the webbing at regular intervals. Feka or Gumao jal of Hoshangabad and Kapa jal of Medagachi belong to this type. The nets may be operated from boats or from shores.

An interesting method of fishing is 'Dhor' practiced with cast net at Hoshangabad. The cast nets, two to three in numbers, are fixed flat with the base of the net forming the mouth. The nets are held in position by bamboo stakes. Fishes are driven to this net by dragging scare lines as in 'Chir' fishing. As soon as the fishes have entered the net the upper portion of the mouth of the net is dropped confining the fishes in the net. The fish captured are mainly *Mystus* spp.

Gill Nets

These are rectangular with lead with or without a foot rope. The presence of sinkers is also optional. The difference between riverine and lacustrine gill nets is only in the size of mesh and the fishing height. The nets with minimum fishing height are noticed in riverine gill nets while lacustrine gill nets have maximum height.

Numerous variations are noticed in the operation of gill nets. In Gochail jal of Allahabad when fishes are gilled, the floats of that particular region sink giving indication to the fishermen, who immediately remove the fish. Against the usual practice of setting the net across the river, kadela of Hoshangabad is set parallel to the shore. Thangadi is payed in the form of an 'L' with the shorter arm of 'L' perpendicular to the shore line. Scaring by beating on the sides of boat is practiced in the operation of this net. The nomadic tribe, Burudebesthas, operate their nets from peculiar saddle–shaped rafts formed by securing dried fruits gourd or tins and paddling these rafts by the leg. Kuto jal of Hoogly Estuary is remarkeble because of the presence of three vertical bamboo spreaders between the head and foot rope. These spreaders serve to keep the net during strong current that exists in the estuary. The crafts used for operation are mostly plank built boats.

Purse Nets

The net has an elliptical or bow–like outline with crescent shaped mouth. The upper and lower lips of the mouth are supported by arc shaped bamboo frames. The length of the are ranges from 5 to 7 meters. The webbing is rectangular in Kamel of Allahabad while it is polygonal in Shangla jal of Hoogly Estuary. The mesh size ranges from 42 to 50 mm bar. The material may be hemp, cotton or even nylon. The closing–cum–hauling rope 'Kassi' is fixed to the middle of the lower arc and passes through a ring fixed to the idle of the upper arc. The length of the rope that can pass through this ring is limited by fixing a knot or seizing along the length of closing rope just above the limit of upper arc. This length corresponds to the actual vertical opening of henet and ranges from 0.85 to 1.5 meters. The feeler rope 'Kutni' is a thin rope cotton or hemp branching into two to four towards the base and fixed on the upper aspect of the net near the lip. In Kamel the feeler rope divides into two while in Shangla jal there are three to four branches. A weight of stone or iron is fixed to the middle of the lower frame where the closing rope is fixed to it. The closing rope is sometimes substituted by slender bamboo rod when operated in shallow waters.

Usually one net is operated from one boat at a time. The net is operated from the fore of the boat. The boat drifts alongwith the current and with the drifting of the boat, the net is lowered in water with the haul rope in one hand and feel rope in the other. When any fish enters the net it cause a jerk which is instantaneously felt by the fisherman through the feeler rope. At Allahabad, usually two nets are operated while in Hoogly Estuary one net is operated from one boat at a time. Catches are migratory hilsa. In the season, it is common sight that the half number of boats are drifting downwards with the current and others are padding up for drifting down again.

Marketing of Fish

It is of interest to note that fish caught from Inland sources are almost entirely diverted fresh to fish markets. The long term preservation techniques are employed only sparingly. Incidentally, it may be noted that 80% of the fish arriving in Calcutta markets; the largest fresh fish consuming city in India, consuming more than 50 tones of fish per day consists of fresh water fish (Nair, 1971). Almost the entire quantity of inland water fish produced in India is marketed fresh. Fish is in great demand in fresh state in market like Calcutta and not affected by marketing problems like seasonal or regional gluts. Nearly 80% of fresh fish carried by the Indian railways, the principle mode of fish transport in internal trade, are fresh water species.

Traditional fishermen are engaged in fishing only. For disposal of their catches, they depend upon their house–hold members or the fish merchants. The fishermen carry small quantities by head loads to nearby places. In some places, retail marketing is entirely carried out by fisher women. Large quantities are auctioned by the fisherman's agents or fish merchants on the landing centers or are transported to the whole sale market. In some states fisheries cooperative federations have been established to develop fish marketing. The central fisheries corporation and some state fisheries corporation also promote marketing. Some of them do not have any link up with production units / cooperatives. An efficient marketing system is very rich needed for better management of the catch and distribution of the same.

The ministry of Agriculture, Directorate of marketing and inspection, published a brochure in 1948 on the marketing of fish in India. In this brochure, scattered fishing centers, primitive method of capture, preservation and transport system have been discussed.

Taking the country as a whole, the breakup of fresh water fish production comprised carps 34.3%, catfish 32.7%, live fish 9.8%, prawns 7.5%, feather backs 4.7%, mullets 9.1%, eels 0.8%, herrings 0.4% and, miscellaneous fishes 3.7%.

The Government of India Ministry of Agriculture vide its letter No.31035/11/78 (T-1) of June, 1980 asked the Indian Institute of Management, Ahmedabad to carry out a study for the Inland fisheries of marketing sector. This study was published ('Inland fish marketing in India) by Indian Institute of Management in year 1985. The study covered the following main parameters of Inland fisheries sector.

1. Infra structural facilities covering the availability of seed
2. Present market structure
3. Distribution system
4. Costs and margins
5. Financial aspects
6. Price-spread from consumer to fishermen
7. Infra structure comprising seed production and its distribution keeping in view, the requirement of fish culture, reservoirs and geographical dispersion.

Fish Markets

Some of the markets served as assembly centers of outstation dispatches and were termed transit and terminal markets. Where the outflows, as percentage of the inflows, were less than half of the market arrivals, such markets were termed terminal markets. The following seven important markets were surveyed.

Sl.No.	Market	Type
1.	Banglore	Terminal
2.	Calcutta	Terminal
3.	Delhi	Transit and Terminal
4.	Hyderabad	Transit and Terminal
5.	Lucknow	Terminal
6.	Madras	Transit and Terminal
7.	Vijayawada	Transit and Terminal

Five out of six largest metropolitan were covered. Bombay, the second largest metropolitan city, was purposely left out as it was mainly a marine fish operation center.

Inland Fish Flows

Among the transit and terminal markets, outflow as percentage of total arrivals was maximum in Vijaywada (72.86%) followed by Madras (60.00%) and Delhi (51.86%). In quantity, Delhi led with 5,586 tones dispatched. Lucknow, though itself a terminal market, dispatched 1,000 tones to other terminal markets. The main terminal market was Calcutta, and all its dispatches were by rail.

Markets Intermediaries

Six principal intermediaries operated in the market (a) Commission agents, (b) wholesalers, (e) wholesaler–cum-commission agents, (d) wholesaler-cum-retailers, (e) retailers and (f) vendors.

Wholesale intermediaries

Commission agents operated at all centers. They extended term loan to their suppliers to ensure supplies and credit to retailers and vendors. The highest number of commission agents were at Calcultta (328) followed by Delhi (70). Madras had 62 exclusive commission agents and 12 commission agents–cum-wholesalers.

Wholesalers mostly contracted water bodies from fisheries departments, co-operatives, or corporations. The pre-harvest contracting of private ponds was also prevalent. All outstation dispatches were on wholesale basis. Wholesalers who handed outstation dispatches mostly bought from commission agents. Delhi had 25 exclusive wholesalers and Lucknow 7. Hyderabad and Madras had 45 and 12 wholesaler-cum-retailers respectively.

Wholesale Markets

The centers surveyed had 7 wholesale markets and 12 wholesale-cum-retail markets. Calcutta had six exclusive markets and Delhi one. Madras had four wholesale-cum-retail markets, Banglore three and Hyderabad twol. Delhi, Lucknow and Vijaywada had only one wholesale-cum-retail market.

The markets handled both marine and inland fish. All the wholesale markets had inadequate truck parking space and poor standards of hygiene. The wholesale market at Delhi was not even approachable by any mean of transport, and all fish movements were on head load up to the nearest main road.

Retail Markets

Retailing was done at 145 markets, which included 12 wholesale-cum-retail markets. Calcutta had the highest number (62) of retail markets; Madras and Delhi had 41 and 25 retail markets. Eighty percent retail markets were of small sizes with less than 50 retailers, and the rest 19 markets were of medium sized. Large and medium–sized markets were only in Calcutta, Madras and Delhi.

The capacities of retail markets at all centers were inadequate to accommodate all operating retailers. The capacity needed to accommodate surplus retailers was the highest in Calcutta with 1,060 excess retailers, followed by Hyderabad with 580, Madras with 528, Delhi with 337 and Lucknow with 275 excess retailers.

Price-spread in Terminal Sale through Retailers

The average consumer price per kilogram was the highest at Calcutta (Rs.16.37), followed by Lucknow (Rs.11.42) and Hyderabad (Rs.9.21) and the lowest at Vijaywada (Rs.7.73), the supplier received the highest share in consumer rupee at Delhi (88.92%) at Calcutta (83.19%) and the lowest share at Madras (58.40%).

The profits percentage of retailers was the highest (28.10%) at Madras, followed by 20.20% at Banglore, and the lowest (3.61%) at Delhi. The commission agents earned a higher share in consumer rupee at Vijaywada 14.20% followed by Madras (11.10%) and Delhi (2.17%).

Price–spread in Terminal Sale through Vendors

The price–spread trend in sale through vendors and retailers was similar. However, the average sale price of vendors was more than retailers sale price: at Calcutta the vendors price was Rs.16.37 per kg. At Vijaywada Rs.7.73 and at Delhi Rs.20.40. The price realized by suppliers was the highest at Delhi (82.21%), followed by Calcutta (82.71%) and Vijaywada (63.10%).

Economics of Retailers

The quantity handled was the highest at Hyderabad (20.80 tones), and the lowest at Calcutta (8 tones). The annual turnover was the highest for Delhi (Rs.2.66 lakhs), followed by Hyderabad (Rs.1.91 lakh), and the lowest for Madras (Rs.0.64 lakh). The daily operating profit at Delhi was the highest at Rs.72.03 followed by Banglore (Rs.69.33) and Vijaywada (Rs.30.77) and the lowest at Madras (Rs.27.37).

Economics of Vendors

The average annual quantity handled varied between 11.64 tones at Hyderabad and 4.51 tones at Madras. It was 10.80 tones at Banglore and 5 tones each at Calcutta and Delhi. Annual of vendors was the highest (Rs.1.02 lakhs) at Calcutta, followed by Hyderabad (Rs.0.92 lakh) and Banglore (0.77 lakh), and the lowest at Madras (0.32 lakh). The daily operation profit was the highest at Banglore (Rs.40.52), followed by Calcutta (Rs.36.85), Lucknow (Rs.36.12), and Vijaywada (Rs.19.36) and the lowest at Madras (Rs.16.74). The vending business at all centers was viable, and each vendor earned much more than the daily wage labour rate.

Fishermen

According to ancient Hindu legislation, fishermen belong to the Sudra or Servile caste. The fishermen belong to an economically and socially backward class and generally ill educated class. The fishermen of the country have a distinct tradition of their own. They belong to all the major religions, namely Hindu, Christian and Islam and to several communities which differ form state to state.

They have adopted varied vocations which mainly connected with water. They are divided into innumerable castes and sub castes. It is common the certain sub castes pursuing as particular profession in one State or one region do not do so in other places. In Northern India alone, they have over 108 sub caste the names of which are given below:

Bhoi, Bathma, Bind, Bhoyar, Bathwar, Bathwa, Bow, Bobes, Bheels, Boat, Bowania, Bonia, Bhor, Choy, Chain, Chhabh or Chhabhi, Dhinwai, Dhuria, Dash, Dass, Dunia, Dheamar, Dulia, Gauyr, Kashatriya, Guriya or Gurya, Gohri, Ghawar, Ghoorum, Gondia, Ghoti, Ghathi, Jeewar, Jaiswai, Jalo, Jheekua, Jheeghat, Jhawa, Jalia, Kahar, Kheer, Kewat, Kherwar Kol, Kol Bool, Khulwat or Khulaunt, Khewar, Keri Herod, Kandaria, Karook, Kapadhai, Kahar, Koi, Karda, Mewasi, Kalikunto, Kaiwart, Kacpadhai, Kaharmor, Kardar, Merimix Kali Kunto, Kaiwart, Kachhwohai, Kawar, Karko, Malah, Mehra, Machhia, Mahar, Mahishi, Malo, Kuriari, Malvi, Manjhi, Machwa, Machandar, Mahir, Noria, Neshad, Newari, Nishad Newari, or Niwari, Nauria, Nadesa Naik, Naganpuria, Nimar, Nawak, Parbalia, Path Wahak, Patniwar, Pawadh, Pari, Phulmali, Prabatia, Rohela, Raikawaar, Sarayar, Suraya, Sandhia, Surgharu, Singhadia, Santari, Signath, Sindha, Teer, Turaha.

The list was first published in 1950. Other than fishing occupation, these people also maintain boats and ferries for inland water transport, cultivate makhana and singhara and make cottage handicrafts. Fishing in general is hereditary profession carried out by them from generation to generation with almost all the members of the family taking part in to a varying degree.

Socio-economic Development of Fishermen Communities in India

One of the most important factors that have influenced the utilization and development of the fishery resources in India, is the socio-economic condition of the fishermen. Fishing is generally considered a low profession in India. On account of poverty, the fishermen are forced to depend on middlemen for the marketing of their produce and naturally the major portion of the profit goes to the middlemen. When the fishing craft and tackle do not belong to the fisherman, he has to give away a good portion of his income as rent of the equipment. Whether the fisherman gets is insufficient for his maintenance and that of his family and so he has to resort to borrowing form the middlemen to tide over the off-season. The middlemen who are fish merchants lay down conditions that would prevent the fishermen from selling their produce in the open market. They have to purchase their domestic as well as fishing requirements from these merchants at exorbitant prices. The fishermen thus become indebted for life and in most cases the indebtedness passed from generation to generation. The low standard of their living conditions, the unhygienic surroundings in which they sold their produce and their poor cultural status resulted in their social isolation. This vicious circle of circumstances has crippled the fishermen community both socially and economically.

Co-operative Movement Among Fisherman

Even during the very early days of fishery development in India, it was realized that the socio-economic advancement of the fishermen is essential for the proper development of the fishing industry. In view of the nature of the economic problems faced by the fishermen, it was obvious that the elimination of middlemen by organizing credit facilities and marketing of produce would go a long way in emancipating the fishermen fold. In some of the maritime States of India where fisheries Department were active, attempts were made to form fishermen's co-operative societies. In many cases, the co-operative societies were not organized after a proper economic survey of the communities and so did not serve the vital needs of the fishermen. Under such circumstances it was found difficult to install the true spirit of co-operation in them and the progress of co-operation movement was very slow and rather discouraging in the beginning. The few societies that were organized functioned as mere credit societies. It was soon realized that the individuals have to reach a minimum stage of social development before they can appreciate the principles of co-operation (Pillay, 1956).

Four Decades of Central Inland Fisheries Research Institute

Before the establishment of the CIFRI, the research and development activities in the inland fisheries sector were scattered and disorganized. The Govt. of India in a memorandum issued in 1943 made knows its intention to set up research station on fisheries. The fish subcommittee of Central Government's policy on agriculture, forestry and fisheries endorsed the above recommendation in 1945. The Central Inland Fisheries Research

Station, thus came into existence on 17 March, 1947 at Calcutta with Late Dr. S.L. Hora, the then Director of Z.S.I. as the Chief Research Officer.

The station was soon shifted to Barrackpore whee it was housed in the temporary hutments of the Ministry of Defence near Palta Water works. Some young scientists who were trained abroad joined the Research Station and worked on several new aspects of inland fisheries. Dr. T.J. Job took over as Chief Research Officer from Dr. Hora in June, 1947 and was, in turn, succeeded by Dr. H.S. Rao in August, 1952. The station was once again shifted to Calcutta as construction work for its own buildings was initiated at the site of hutments. Dr. B.S. Bhmachar took over as Chief Research Officer in July, 1954 and was designated as the Director in 1961. The status of the station was also raised to that of an 'Institute'. The Institute came to be located in its own buildings at Barackpore in June, 1959. The buildings house the laboratories, auditorium, aquarium, stores, hostel, guesthouse and residential quarters on a 5.2 ha plot. The onus of Directorship fell on the shoulders of Dr. B.S. Bhmachar in July 1966.

The Govt. of India took a decision to hand over this institute and other fishery institutes to the Indian Council of Agriculture Research on October, 1967. Since, then the institute has played a great role in the development and expansion of fisheries in India.

Research Programmes

The research programme of the institute was oriented in 1968 initially into 20 well-defined problem oriented projects, which increased to 29. These projects cover a number of shorts and long range specific research problems to bring about culture fisheries development and increased fish production. The 29 projects are listed below.

Project 1 Optimum per hectare production of fry, fingerings and fish in culture fishery operations.

Project 2 Induced fish breeding

Project 3 Reservoir fisheries

Project 4 Riverine carp spawn prospecting and collection technique

Project 5 Brackish water farming

Project 6	Freshwater prawn culture
Project 7	Murrel and live fish culture
Project 8	Estuarine and brackish water lake fisheries
Project 9	Selective breeding and hybridization
Project 10	Fish Farm Designing
Project 11	Economics in Fishery investigations
Project 12	Exotic fish culture
Project 13	Cold water fish culture
Project 14	Riverine and estuarine fish catch statistics
Project 15	Fish Pathology
Project 16	Weed control
Project 17	Frog farming
Project 18	Sewage–fed fisheries
Project 19	Hilsa fisheries
Project 20	Water pollution investigations
Project 21	Fisheries of river basins
Project 22	Fish culture in running water
Project 23	Bundh breeding
Project 24	Fresh water aquaculture in urban areas
Project 25	Beel fisheries
Project 26	"All India" Coordinated Research Project on composite fish culture and fish seed production
Project 27	"All India" Coordinated Research Project on Air breathing fish culture.
Project 28	"All India" Coordinated Research Project on Ecology and Fisheries of freshwater.
Project 29	CIFRI, IDRC Project on Rural Aquaculture

The setting up of Central Inland Fisheries Research Institute in 1947 was a landmark in the field of research on breeding and culture technology for the most widely prevalent species of carp. The development of culture practices for other fresh water fishes together with estuarine species of prawns and fish was also carried on significantly by this institute. The Central Marine Fisheries Research Institute on the other hand, played a significant role by giving fillip to and accelerating studies on the cultural aspects of marine prawns, mollusks and fishes in the early seventies

It would be expedient and salutary to categorize fisheries research and development in two phases primary emphasis was placed on evolving and developing fish for seed production. The second phase was sharply distinguished by concentrating on infrastructural and development support to the technology already developed. The seventies are thus, remarkable for a major breakthrough in the development of area specific technology.

In a frantic bid to boost up the technology of freshwater aquaculture, the Indian Council of Agricultural Research established the Freshwater Aquaculture Research and Training Center in 1980. Equipped with modern research laboratories it succeeded in upgrading technical competence and initiating in depth studies to optimize fish production. The center provided a clear understanding of the subjects like nutrient dynamics in tropical pond environment, nutrition and diseases of cultivable species and genetic improvement for producing strains of carps with better cultural research in India, as its contribution to a disciplined research thrust in this area was substantial. The up gradation of the center into the Central Institute of Freshwater Aquaculture (CIFA) in 1985 was a measure of its distinctive achievement. The development of seed and culture technologies of marine species of fish including shrimps and mallows took place precisely because of the work on brackishwater farming by the CIFRI and CMFRI. Thus the Central Institute of Brackishwater Acquaculture (CIBA) came into being in 1985. Concurrently, the Government of India, concentrated on a two told activity, *i.e.* provision of credit and transfer of technology–steps which were oriented towards improved production of seed fish and fish.

Carp Culture

Marked and substantial improvement in the survival of naturally procured carp hatchlings in nurseries was achieved which tended to be meager and slight (not exceeding ten to fifteen per cent) during early sixties at the pond culture station of CIFRI at Cuttack. Fertilizer and maturing application schedules based on the nutrient and physicochemical status of water and pond soil were the off–shots of concerted research. The phenomenal development in modes of treatment for control of aquatic insects in nurseries as also in removal of unwanted weeds resulted in the overall development of nursery technology with 80% survival of fish embryos.

The early sixties were marked by a significant achievement in Induced Breeding (IB) of Indian major carps, such as catla (*Catla catla*), rohu (*Labeo rohita*) and mrigal (*Cirrhinus mrigala*) by pituitary harmone administration to both male and female specimens showing ripe gonads. Within a brief span of time, this technology came into operation pervasively for all the carps including the exotic silver carp and the grass carp. Another significant achievement was the development of composite fish culture technology. Based on the six ecologically compatible carps–three Indian major ones and three exotic *i.e.* silver, grass and common carps, this carp polyculture technology proved to be quite high yielding. The different practices employed for this purpose consisted of fertilizing, manuring and supplementary feeding schedules as well as removal of unwanted weeds and fishes.

The all India Co-oriented Research Project (AICRP) on Composite fish culture (CFC) and fish seed production (FSP) came into existence in the year 1971 as a result of remarkable development of carp breeding and culture technology in research farm of CIFRI with the sole purpose of locating natural carp seed procurement grounds and augmenting carp seed supply in the country. Subsequently it was merged into the All India Co-oriented Research on Composite Fish Culture (CFC) and fish seed production (FSP).

A major break through by the AICRP on CFC and FSP was perceptible in the initiation of adoptive research together with the development and demonstration of area specific CFC technology. All this resulted in the optimum production of 3 to 10 tones per hectare in nine to twenty months as well as breeding of farm–reared carps at all the centers in the states of West Bengal, Orissa, Andhara Pradesh and Tamil Nadu. The viability of technology in the farmer's pond was further demonstrated under the rural aquaculture projects sponsored by the ICARI-ADRC in Orissa and West Bengal.

Research on the breeding biology of carps was carried out by the co-oriented Research Project which demonstrated the theory of gonadal hydration in the successful breeding of silver and grass carps in bundhs. Consequently, a large number of bundhs were established. Basically, they simulated the riparian environment with heavy freshwater discharge in West Bengal, where fish could be bred easily.

In the early seventies a glass jar hatchery for carps was developed by CIFRI and close on the heels came variants of water circulating hatcheries of multiple designs and capacities. The Central Institute of Fisheries Education (CIFE) Bombay contributed significantly to the growth of a new and more effective hatchery system. On the top of all, methods were evolved to induce the major carps to breed twice within the same season by taking into consideration a better brood stock management. The sustained efforts succeeded in boosting up seed production. In fact, the phenomenal rise in carp breeding was accelerated by multiple spawning and prolongation of breeding season from two to seven months.

The eighties were marked by a perceptible advance in technology development together with the proliferation of fish farmers training centers and extension support to Fisheries Institute. While a World Bank–aided project gave birth to modern commercial hatcheries, the National Hatcheries project boosted up seed production of carps.

In 1973, a vigorous programme of Fish Farmers' development Agencies at the district level was launched to ensure administrative and infrastructural support. The overall impact of this programme was reflected in the training of beneficiaries, mobilization of inputs, extension support to fish farmers and arrangement of institutional finance through bank credits. While Fish Farmers' development Agencies put in concerted efforts to utilize the fallow ponds resources, the productivity of both seed and fish shot up as a result of constituent efforts by traditional farmers of carp culture ponds.

The phenomenal increase in the number of Fish Farmers' Development Agencies ranged from fifty covering 1047 hectares in 1979–1980 to two hundred covering 1,85,000 hectares in 1987–1988. This had a visible impact on the average productivity, which mounted up from 582 kg/hectare/year in 1979–1980 to 1556 per hectare per year in 1988 in the FFDA ponds. A tremendous increase in the production of fry and fingerlings took place from 1048 million in 1979–1980 to 9300 million in 1988. Simultaneously, the inland fish production also rose from 0.89 million tones in 1980–1981 to 1.38 million tones in 1987–1988 (Sinha, 1991). The rise in the inland fish production could be attributed largely to fresh–water aquaculture precisely because general degradation and water management had led to a marked decline in riverine and reservoir fisheries development. A sizeable number of fresh water fish farms sprang up in coastal Andhra Pradesh as a direct off shoot of the training imparted by the CIFE and backed up by extension support. The annual production from this region was estimated at 60,000 tones.

Carp Farming

Accumulation of large deposits of silt in a number of non-drainable ponds meant for carp culture posed a major hurdle in carp culture. It is well known in the ponds associated with the largest human and animal livestock population as also the ponds with Microcystic bloom contain the highest number of bacterioplankton. The Microcystic bloom which brings up the sediment nutrient in the early stages as well as during their vertical movement is commonly found in ponds with apparently deficient water column but with thick, nutrient rich and anaerobic sediment. The decay of the Microcystic bloom the water column releases nutrients which become available for other algal species. Efforts were, therefore, made to evolve suitable methods of recycling the nutrients from the pond bottom in order to make the ponds more productive. The non-availability of the fish toxicant, *i.e.* Mahua oil–cake commonly used for eradicating unwarranted fishes spurred on frantic search for alternatives. Calcium hypochlorite (CaOcl) either alone or in urea yielded promising results.

In a large number of nursery ponds as also in some rearing and stocking ponds supplemental feeding with rice and oilcake was done. Some feed formulations for carp brood stock and seed were also accomplished. Experiments with leaf proteins and unconventional feeds were also taken into consideration. The issues relating to the availability of potential feed ingredients, their nutritional value and costs were also explored.

The CIFA embarked on systematic research into fish diseases with the consequence that many bacteria were identified and isolated. Investigated into a virus associated with the Epidermal Ulcerative Syndrome (EUS) of fish was taken up at the CIFA. The Central Inland Capture Fisheries Research Institute (CICFRI) launched a collaborative research programme on Net-work of Aquaculture in Asia (NACA) under FAO/UNDP project.

The CIFRI did a magnificent job by acquiring device for converting liquid ammonia into manure for the control of aquatic weeds. Subsequently, this technique was tested widely, investigation into the nutrient value of weeds and the modes of incorporating them as sources of energy and protein in fish feeds was conducted with promising results. A substantial increase in the survival of as much as 80 percent and stocking rates up to 10 million hectares into nurseries was made possible by concerted research. Addition of cobalt chloride (0.01 mg/fish/day) showed high survival rates. The survival rate was improved by treating the pond with soft organophosphate prior to the stocking of hatchlings in the nursery ponds. Adequate aeration and provision of formulated feeds results in raising 3000–5000 spawn/m with 90 percent survival at the Central Institute of Fisheries Education (CIFE), Bombay.

Experimentation in cage culture of carps yielded tangible results. By applying feed which was mixture of silkworm pupil powder, rice bran and groundnut oil-cake with wheat flour as a binding material, a cage of $10.5m^2$ could produce 140 kg of fish in six months with a stocking rate of 400 per cage.

In sewage ponds, the production rate of 7200 kg per hectare per year at the Rahara Farm, Kharda (W.B.) provided the abundant scope for culture of silver carp with Indian major carps. An yield of 9350 kg per hectare per year was indicated by culture of chromis mossambicus in sewage fed ponds without supplementary feed with sewage, although without any other feed or fertilizer, an average yield of 7 tones of major carps of Macrobrachiam rosenbergii per hectare per eight months was achieved.

For carp species a good deal of effort was made to practice integrated culture of fish with paddy or animal such as ducks and pigs. The culture of air-breathing fish with paddy cultivation was also explored successfully. The integrated paddy-cum-fish culture resulted into an yield of 700 kg of carp per hectare per year. While the production from fish cum pig culture was estimated at 700 kg of carps per hectare per year, it was 4223 kg of carps per hectare per year form duck-cum-fish culture.

Air-breathing Fish Culture

The air-breathing fishes, *i.e.* Clarias batrachus, Heteropneustes fossilis, Anabas testudineus, Channa marulius, C. striatus and C.

punctatus are ideally suited for cultivation in weed-chocked water area with low content of dissolved oxygen. Elaborate and detailed investigations into the development of cultural practices for optimum production of air-breathing fish under different management system in the states of Karnataka, Andhra Pradesh, Assam, West Bengal and Bihar were launched by the all India Co-oriented Research project on Air-breathing Fish Culture (AUCRO on ABF) in 1971. A substantial yield of over a tones per hectare per eight months was achieved in mixed culture of Clarias, Heteropneustes, Anabas. With high density of stocking, intensive feeding and water replenishments, Clarias was found to yield 50 tones per hectare per three months. However, non-availability of seed and constraints of feeds accounted for the low input of the air-breathing fish culture enterprise.

Hilsa Fisheries

The construction of weirs and barrages on rivers in the lower stretches on a large scale led to a sharp decline in the yield of anadromous Hilsa ilisha. IN 1986, the CIFRI achieved success in the breeding of procured mature specimens of Hilsa through stripling. Since then it became a recurrent feature of breeding of Hilsa by this method. In the Ukai reservoir on River Tapti in Gujrat artificially bred Hilsa demonstrated their potential for growth. An analysis of cultural practices revealed the size of Hilsa ranging between 325 and 365 mm in a couple of years.

Cold Water fisheries

There are abundant and extensive resources in this country for the development of temperate aquaculture. The breeding, seed rearing and culture of the rainbow trout, Salmo trutta received impetus from sport culture fisheries in hill streams for the tourist industry. In Kashmir, the proliferation of hatcheries occurred with the consequent ushering in the commercial culture of the rainbow trout. For the wide expansion of the technology on an extensive scale, difficulties arose due to shortage of high protein feed together with the soaring cost of fish meal. It would be pertinent to state that the mahaseers (Tor spp) have been the legendary sport fish of India with high table value. Tor khudree was successfully bred in private sector farms at Lonavala near Bombay. The standardization of breeding technology for T. putitora, the most prized specimen of temperate water, was achieved in the foothills of the Himalayas. However, the seed rearing technology and commercialized feed required refinement to promote their culture on a large scale. The needed research back-up was provided by the National Research Central on Cold Water Fisheries by the ICAR.

Reservoir Fisheries

AICRP on Ecology and Fisheries of Reservoirs (EFR) was launched in 1971 at the CIFRI. By identification of productive ecological potentials together with formulation of stocking, harvesting and management policies, the scientific basis of aquaculture in fresh water reservoirs was explored. But the implementation of fishing policy remained elusive and proper stocking could rarely take place in such reservoirs. The overall impact of research on boosting up productivity was not adequate. The rise in productivity in small reservoirs, stocked and managed by the CIFRI, ranged from about 15 kg per hectare per year to over 100 kg per hectare per year.

Manpower Development

The ICAR initiated a number of steps to pave the way for eminently qualified and trained manpower to execute the research programmes like Aquaculture Engineering, Nutrition and Feed formulation, Genetic improvement of stocks, Fish health management, Production and Marketing economics and Controlled breeding of species. A major break through in this field was achieved by elevating the Central Institute Fisheries Education Mumbai (CIFE) to the status of Deemed University for Fisheries. Stress on research in basic and applied aspects of fisheries remained the focal point of the academic programmes at the postgraduate level in the CIFE. The setting up of the following thirteen laboratories was a measure of the upgradation of research facilities (Sinha, 1991).

1. Fishery oceanography and EEZ fisheries management.
2. Soil, water and productivity management.
3. Environmental monitoring, pollution and diseases.
4. Reproductive physiology, genetics and breeding.
5. Fish digestive physiology, nutrition and feed formulation.
6. Tropical and Temperate acquculture.

7. Mixed and Integrated fish farming.
8. Fishing, Processing and Product Development.
9. Fishery machinery, instruments and vessels management.
10. Fisheries–economics and Fisheries' socio-economics.
11. Fisheries' Resource Dynamics and Project Planning.
12. Fisheries Extension and Transfer of technology.
13. Bio-technology and Hi-tech systems.

Most of the above mentioned laboratories are preoccupied with training manpower and carrying on research at the highest level in order to meet the needs of qualified manpower for aquaculture developments. These departments in the CIFE as well as in other Fisheries' institutes would play a significant role in accelerating the development of aquaculture in India.

Chapter 6

"INDIAN FISHERY: A Future Vision"

Indian Fisheries have made great strides during the past five decades. As a result Indian at present produces over 63.9 lakh tones of fish and shell-fish per year from all resources. This production over eight fold from 7.5 lakh tones production during 1950. India ranks 6th in World total fish production and second in inland fish production. Fisheries Sector contributes about Rs. 22,200 crores to the Gross Domestic Product (GDP) which is about 1.4 percent. the total GDP and 4 percent of the contribution of the Agriculture sector. The share of inland fisheries sector in the total production which was about 29 percent during 1950–1951 has now gone up by over 54 percent.

The Aquaculture sector has shown overwhelming growth of 468 percent during the last two decades *i.e.* 3.7 lakh tonne during 1980 to over 22 lakh tones in 2002. the freshwater aquaculture continues to make major contribution *i.e.* 95 percent. The three Indian major carps contribute production of the order 16 lakh tones which contributes a major component of the annual culture production over 22 lakh tones as in 2002. The fisheries have been an important source of foreign exchange for the Country and earnings have crossed Rs. 5000 crores during 2000–2001. It also have great role in socio-economic development of the country. It is a large employment generator as it stimulates the growth of a number of ancillary industries and is major source of livelihood living in coastal area. More than 70 lakh people are fully dependent on fishing and almost equal number are dependent on the allied activities for their livelihood.

The annual growth rate is 6 percent in the last decade which is higher than the most of the food producing sectors.

International declarations have highlighted the importance of aquaculture. Kyoto Declaration in 1976 states that aquaculture has made significant contributions to food production and means for revitalizing rural life and producing high value food. It could contribute to enhancement and maintenance of wild stocks and provides an efficient means of recycling and upgrading low-grade food and waste into high-grade protein rich food. Aquaculture, with high levels of integration with other farming systems, merits full support as part of comprehensive land and water use. The Bangkok Declaration and strategy of 2000 emphasizes on development, adoption and application of environmental, economic and social sustainability assessment criteria and indicators of aquaculture development, improved management practices, codes of good practices for aquaculture efficient use of resources and promotion of aquaculture as a means of improving environmental quality and resource use.

Fisheries and Aquaculture in India have witnessed and impressive transformation from a highly traditional activity to one based on a well developed and diversified infrastructure with immense potential for industrialization. Aquaculture accounts for about one million tonne *i.e.* 58% of Inland fish production. Aquaculture would account for more than 70% production of Inland sector *i.e.* 6.39 million tonne as per statistics for the year 2003–2004. Inland production is 3.5 million tonne where as Marine production is 2.9 million tonne. The details are given in the following tables:

Sl.No.	*States/UTs*	*2000–01*	*2001–02*	*2002–03*	*2003–04**
(A)	**INLAND**				**(in tonnes)**
1	Andhra Pradesh	4,07,80	4,71,165	5,79,403	6,80,708
2	Arunachal Pradesh	2,500	2,600	2,600	2,652
3	Assam	1,58,620	1,61,450	1,65,521	1,81,000
4	Bihar	2,22,160	2,40,400	2,61,000	2,66,490

Sl.No.	States/UTs	2000–01	2001–02	2002–03	2003–04*
5	Goa	4,240	3,358	4,247	3,600
6	Gujarat	40,261	50,774	34,267	45,479
7	Haryana	33,040	34,566	35,182	39,133
8	Himachal Pradesh	7,020	7,215	7,244	6,526
9	Jammu & Kashmir	17,510	18,850	19,750	19,750
10	Karnataka	1,27,468	1,21,196	86,263	70,000
11	Kerala	85,234	78,039	75,036	76,179
12	Madhya Pradesh	48,844	47,457	42,168	50,818
13	Mahrashtra	1,23,266	22,785	1,27,236	1,25,120
14	Manipur	16,050	16,450	16,600	17,600
15	Meghalaya	6,179	4,968	5,372	5,147
16	Mizoram	2,860	3,147	3,250	3,380
17	Nagaland	5,500	5,200	5,500	5,560
18	Orissa	1,38,556	1,68,056	1,74,201	1,90,017
19	Punjab	52,000	58,000	59,475	83,650
20	Rajasthan	12,121	14,269	13,400	14,300
21	Sikkim	140	140	140	140
22	Tamil Nadu	1,13,560	1,14,000	1,02,217	1,01,142
23	Tripura	29,420	29,450	16,155	17,980
24	Uttar Pradesh	2,08,286	2,25,371	2,49,837	2,67,000
25	West Bengal	8,79,230	9,15,800	9,38,500	9,88,000
26	A & N Islands	66	61	74	90
27	Chandigarh	82	44	84	84
28	Dadra & Nagar Haveli	43	55	46	54
29	Daman & Diu	0	0	0	0
30	Delhi	3,980	3.2	2,250	2,100
31	Lakshadweep	—	-	-	-
32	Pondicherry	4,350	4,900	4,900	5,200
33	Chhatisgarh	43,386	95,763	99,801	1,11,050
34	Uttaranchal	9,074	6,422	2,252	2,563
35	Jharkhand	42,600	1,01,000	45,380	75,380
	Total (A)	**28,44,832**	**31,26,163**	**31,78,651**	**34,57,892**

Sl.No.	States/UTs	2000–01	2001–02	2002–03	2003–04*
(B)	MARINE				(in tonnes)
1	Andhra Pradesh	1,82,502	2,04,940	2,42,495	2,63,926
2	Goa	67,328	66,550	72,287	83,756
3	Gujarat	6,20,474	6,50,829	7,43,638	6,09,136
4	Kamataka	1,75,906	1,28,415	1,80,161	1,87,000
5	Kerala	5,66,571	5,93,783	6,03,286	6,08,525
6	Maharashtra	4,02,838	6,14,268	3,86,860	4,20,017
7	Orissa	1,21,085	1,13,893	1,15,008	1,16,880
8	Tamil Nadu	3,67,855	3,70,998	3,79,214	3,73,000
9	West Bengal	1,81,000	1,84,300	1,81,500	1,81,600
10	A & N Islands	27,618	27,021	28,228	31,058
11	Daman & Diu	16,382	21,524	11,258	13,766
12	Lakshadweep	12,000	13,650	7,496	10,031
13	Pondicherry	38,950	39,600	39,600	42,800
	Total (B)	**27,80,510**	**28,29,771**	**29,91,031**	**29,41,495**

Keeping in view the immense potential and prospects of Fisheries and Aquaculture Development in the Country National Policy an Aquaculture Development has been formulated.

Objectives	*Priorities*
☆ To evolve economical aquaculture systems and integration of aquaculture with agriculture, horticulture and livestock.	☆ Integration of aquaculture with agriculture, horticulture, animal husbandry, etc.
☆ To evolve culture technologies suitable to Indian conditions for coldwater fish.	☆ Socio-economic studies on aquaculture and its integration with agricultural activities and economic evaluation of various aquaculture technologies.
☆ To evolve semi-intensive and intensive aquaculture technologies for prawn.	☆ Development of semi-intensive culture technologies for brackishwater and freshwater prawn
☆ To develop prawn hatcheries for both freshwater and marine prawns for the purpose of raising seed.	☆ Technologies for seed production and culture of marine and freshwater crustaceans.

Objectives	Priorities
	☆ Aquaculture engineering, specially hatchery construction, pond, pen and cage construction/ management of large water bodies.
☆ To evolve economical culture systems (pond, pen and cage) for marine and brackishwater fish, including seed production of such species in hatcheries.	☆ Establishment of hatcheries. ☆ Maturation and seed production of brackishwater and marine fish, carp, mollusks, ☆ Development technologies for seed production and culture of mollusks, including bivalves.
☆ To evolve semi-intensive and intensive culture technologies for seaweed of commercial importance and to culture fish food organisms.	☆ Culture and harvesting technologies for seaweed. ☆ Nutrition and feed formulations for cultivable fish species, shellfish and other aquatic organisms of cultural value.
☆ To achieve basic and applied research solutions to problems in fish and shellfish physiology, nutrition genetics, pathology, microbiology, pond and reservoir, lake, cage and pen environments.	☆ Development of basic research facilities for work on various technical aspects.
☆ To undertake selective breeding of rohu, catla and mass produce fry and fingerlings of silver carp and common carp.	☆ Semi-intensive and intensive carp culture, with production target of 15–25 tonnes/ha/yr. ☆ Intensive carp culture in running waters ☆ Semi-intensive culture of cat fish.
☆ To evolve and commercialize culture technologies of oyster producing freshwater pearl and marine pearl.	☆ Development of culture technology for marine oysters and pearl culture. ☆ Development of culture technologies for freshwater mussels and freshwater pearl culture.
☆ To transfer technologies to users	☆ Extension, teaching and training of farmers, extension workers, teachers, etc.
☆ To train farmers, trainers	

The strategies used for implementation of the National Aquaculture Development Plans are given below:

1. Enhancing production and productivity of fish farmers.
2. Enhancing total aquaculture production by developing culture technologies of finfish, shellfish and Seaweed, etc. for freshwater, brackishwater and seawater resources of the country.
3. Development of freshwater and marine pearl culture and culture of freshwater mussels, marine oysters, bivalves and sea cucumber.
4. Development of infrastructure training and demonstration for dev elopement of human resources for aquaculture, including integrated aquaculture.
5. Socio-economic studies in aquaculture and diversification of culturable species and standardization of various aquaculture systems with maximum economic profitability, sustainability and environmental stability.
6. Selective breeding of rohu, catla, silver carp and common carp.
7. Development of efficient and economical fish and prawn feeds.
8. Disease prevention and control.

The goals and objectives of the National Aquaculture Research Sector are given below:-

1. To evolve aquaculture systems that are economical, sustainable and eco-friendly, and integrate aquaculture with agriculture, horticulture, livestock and poultry.
2. To increase production and to evolve culture technologies suitable to Indian conditions.
3. To evolve semi-intensive and intensive aquaculture technologies for prawns.
4. To evolve economical culture systems (ponds, pen, cage_ of marine and brackishwater fishes, and produce seed of these fishes through hatchery management.
5. To evolve semi-intensive and intensive technologies for seaweeds of commercial importance and for culture of fish food organisms.
6. To evolve and commercialize culture technologies of freshwater pearl and marine pearl producing oysters; and

The Action Plan for the implementation of National Aquaculture Development Plan is given below:

Action Plans	*Objectives*	*Target Beneficiaries*	*Expected Results*
Freshwater carp culture	☆ To increase productivity and production of carp through culture systems ☆ To integrate carp culture with agriculture, horticulture, animal husbandry, etc ☆ To extend carp culture to various climatic zones through the Fish Farmers Development Agencies	☆ Carp farmers ☆ Carp hatchery managers and seed producers ☆ Government sector	☆ Carp productivity above 2–7 t/ha/yr ☆ Carp productivity in intensive culture of 15–25 t/ha/yr ☆ Extension of carp culture to more than 2254 lakh ha
Culture of catfish and air-breathing finfishes	☆ To develop extensive and semi-intensive culture technologies ☆ To develop and standardize induced breeding and larval rearing technologies ☆ To integrate air-breathing fish culture and paddy culture	☆ Catfish and air breathing fish farmers ☆ Entrepreneurs ☆ State Fisheries Departments ☆ Aqua culturists	☆ Production rates of 5–10 t/ha/yr ☆ Running water and water exchange rates of 30–100 t/ha/yr ☆ About 1 million larvae ☆ Integrated of paddy cum fish culture technology package

Contd...

Action Plans	Objectives	Target Beneficiaries	Expected Results
Breeding hatchery rearing and culture of prawns and brackishwater fish	☆ To develop standard technology for brooder-raising for brackishwater and freshwater prawns	☆ Prawn farmers ☆ Brackishwater Farmers Development Agencies (BFDAs) ☆ State fisheries departments	☆ Technologies for brood stock and larval rearing and quality seed production through breeding in captivity ☆ Standard culture technologies for prawns and brackishwater fish ☆ Development of hatcheries
Fish Seed and hatchery development	☆ To establish fish seed hatcheries ☆ To ensure the availability of quality fish seed to fish farmers	☆ Fish farmers ☆ Fish hatchery operators ☆ State Fisheries Departments ☆ Private and cooperative seed producers	☆ Establishment of 47 commercial fish seed farms and hatcheries in the country
Integration of aquaculture livestock and poultry	☆ To develop unified farming systems integrating aquaculture with agriculture, livestock, poultry, horticulture	☆ Aquaculturists ☆ Agriculturist ☆ Entrepreneurs	☆ Development of an integrated aquaculture-agriculture-livestock technology
Development of recreational and coldwater fisheries	☆ To develop breeding, larval rearing and culture technologies for coldwater fish and ornamental fish	☆ Coldwater area fish farmers ☆ Private sector (entrepreneurs) ☆ State Fisheries Departments	☆ Standard economically viable technologies for breeding, seed raising and culture of coldwater and ornamental fish

Contd...

Action Plans	*Objectives*	*Target Beneficiaries*	*Expected Results*
Fisheries development of floodplain lakes	☆ To develop about 2.0 lakh ha of floodplain lakes for culture and capture fisheries	☆ State Fisheries Departments ☆ Fisheries cooperatives	☆ Standardized package of practices for culture in floodplain lakes to increase inland fish production
Cage culture of fish in reservoirs	☆ To develop economically viable technology for cage culture in reservoirs	☆ State Fisheries Departments ☆ Fisheries cooperatives	☆ Development of package of practices for cage culture of important fishes in reservoirs
Human resources development	☆ To develop trained and skilled fisheries personnel	☆ Fisheries Departments ☆ Fisheries colleges ☆ Fisheries institutions ☆ State Agricultural Universities	☆ Conduct of refresher courses for fishery workers, demonstrators, teachers, extension workers and researchers

The main issues for formulating and implementing the National Aquaculture Development Plan are given below:

Category	*Issues*	*Constraints*	*Required Actions*
Administrative and institutional aspects	☆ Lack of sufficient infrastructure ☆ Weak interagency & institutional linkages	☆ Lack of funds ☆ Weak institutional linkages	☆ Budgetary allocation for infrastructure ☆ Strengthen linkages and improve coordination
Legal aspects	☆ Conflicts between aquaculture and agriculture on environment	☆ Lack of clear guidelines for aquaculture farm and area development	☆ Legal guidelines on aquaculture areas/farms
Information	☆ Weak technology transfer ☆ Lack of technology packages	☆ Lack of funds for publication production and distribution ☆ Lack of awareness on aquaculture technologies	☆ Strengthen and expand extension services ☆ Develop technology packages ☆ Increased budgetary allocation for extension
Human resources	☆ Lack of skilled extension workers at block and village levels ☆ Lack of training courses demonstrations ☆ Lack of coordination among agencies and institutions	☆ Lack of funds ☆ Lack of information on actual human resources requirements for aquaculture development of a country	☆ Increased budgetary allocation ☆ Increased number of courses, training of courses, training programs, and fellowships ☆ Coordinated training & HPO programmes

Contd...

Category	*Issues*	*Constraints*	*Required Actions*
Technical aspects	☆ Lack of information on feed ingredients, feeds & feed mills ☆ Lack of facilities for disease diagnostics, feed and seed production, & processing	☆ Improper processing, storage and distribution of feeds cost wise and efficiency wise	☆ Establishment of feed mills, hatcheries, post harvest processing & storage plants, disease diagnostic facilities
Physical and environment aspects	☆ Identification of further aquaculture sites in fresh, brackish and marine waters ☆ Lack of pollution, water and mineral seepage monitoring	☆ Lack of survey data on water bodies ☆ Uncontrolled use of fertilizers and feeds ☆ Pollution ☆ Unplanned development of aquaculture farms	☆ Survey & Assessment of Potentials of Water Bodies ☆ Control and monitoring aquatic pollution
Socio-economic aspects	☆ Difficulty in obtaining credits & loans ☆ Lack of facilities for marketing ☆ Inadequate facilities for processing and marketing	☆ Funding agencies are not well informed of potentials of aquaculture technologies ☆ Inadequate marketing channels	☆ Increased access to credits & support of aquaculture by funding agencies ☆ Development marketing channels

to develop technologies for freshwater mussel and mollusk culture and marine bivalve culture.

7. To develop solutions to various technical problems which solutions would lead to economical and efficient aquaculture.
8. To transfer aquaculture technologies to users, train farmers, trainers and teachers; and develop human resources for aquaculture.

The future vision for the development of fisheries in India requires important steps. Deep sea fishing policy would provide for exploitation of deep sea resources should be increased by 15%. Advance technology including for fish finding which can help the fishermen requires to be strengthened. The brackish water aquaculture requires to be increased by 5–6 times from the present level of 1.4 lakh hect. to 11.9 lakh hect. Freshwater aquaculture needs to be popularized and the average productivity requires to be increased from 2260 kg per hect. to 3000 kg. per hect. Diagnostic facilities for fish diseases require to be strengthened. The used of bio-technology techniques would increase the fish production. The quality standard for fish and fish production is required to be framed for development of Inland and external trades. The extension mechanism need to be strengthened so that results of technology development should reach the beneficiaries without delay. GIS applications are requires for identification of resources and infrastructure for fishery development. The welfare any safety measures for fishermen requires to be strengthened for proper exploitation and marketing. Appropriate law wherever necessary for protection and conservation resources should be framed. Close cooperation with the developing country which has the common boundaries of sea must be managed cordially.

"BIBLIOGRAPHY"

Ahmad, N. (1943). Fauna of Lahore and fisheries of Lahore. *Bull. Deptt. of Zoology*, Punjab Univ., 1: 253–374.

Ali, Salim A. (1927). The Mughal Emperors of India as Naturalists and sports man. *J.B.N.H.S.*, 32(2): 268–269.

Alikunhi, K.H. and Chaudhuri, H. (1951). On the occurrence of *Labeo rohita* to the Godavari river system. *Sci. and Cult.*, 16(11): 527.

Alikunhi, K.H. and Chaudhuri, H. and Ramachandran, V. (1951). Response to transplantation of fishes in India with special reference to conditions of existence of carp Fry. *Symposium on Transplantation of Fishes Indo-pacific Fisheries Council 3rd Meeting, Chennai.*

Annadale, N. and Hora, S.L. (1925). The freshwater fish of the Anandaman Island. *Rec. Ind. Museum.*, 27(2): 33–41.

Awati, P.R. and Bal, O.W. (1933–34). *J. Univ. Bombay*, I&II, 1933–34.

Awati, P.R. and Pinto, M.F. (1937). *J. Univ. Bombay*, V, 1937.

Barua, B.M. (1941). *Asoka's Inscriptions*, 2: 97.

Barua, B.M. (1943). *Inscriptions of Asoka*, Part II: Translation, Glossary and General Index, p. 358–374, University of Kolkata.

Bhatti, Hamid Khan and Latief, A. (1951). Catching and marketing off fish in Punjab (The Board of Economic Enquiry Punjab Publication No. 99.

Bhimachar, B.C. (1932–1934). *Half Yearly Mysore University Journal*, VI.

Bhimachar, B.S. and Rau, A.S. (1941). The fishes of Mysore State. *J. of Mysore University*, 1: 141–153.

Bhimachar, B.S. (1942). Report on a survey of the fisheries of the Mysore state. *Fish. Bull.*, 1: 1–39. .

Bose, N.K. (1954). *Man of India*, 34(4): 285.

Brandt, A. Von (1958, 1964 and 1972). *Fish Catching Methods of the World*. Fishing News (Books) Ltd., London.

Breder (1926). *Zoologica*, 4: 159.

Buhler, G. (1894). *Epigraphica Indica*, 11: 258–259.

Chauhan, B.S. (1947). Fish and Fisheries of the Patna State (Orissa). *R.I.M.* 45: 267–282.

Clark, J.G.O. (1948). The development of fishing in pre historic Europe. *The Antiquaries J.*, 28 (Jan–April), London.

Codringtion, K.de D. (1946). Notes on the Indian Mahseers. *J.B.N.H.S.*, 43: 336–344.

Das, K.N. (1939). On a collection of fish from Hazaribagh District, Bihar. *R.I.M.*, 41(4): 437–450.

Day, Francis (1883). *International Fisheries Exhibition Indian Fish and Fishing, London.*

De, Kiran Chandra (1910). Report on the fisheries of Eastern Bengal and Assam, p. 1–77.

Edye, E.H.H. (1923). Report on the fisheries of The United Provinces, Allahabad, p. 1–36.

Fraser, A.G.I. (1942a). Fish of the Poona, *J.B.N.H.S.*, 43(1).

Gopinath, K. (1942). Acclimatisation of foreign fish in Travancore. *J.B.N.H.S.*, Bombay, 43: 267–271.

Grison, W.V. (1938). The meria gonds of Bastar State, London.

Hargraves, H. (1929). Excavations in Baluchistan in 1925, Sampur mound, Mastung and Shor Damb, Nat. Mem. Archaeol. Surv., India. 35: 45–55.

Hill A.V. (1945). *Scientific Research in India*. Govt. of India Press, p. 1–4.

Hora, S.L. (1926a). On the manuscript drawing of fish in the library of the Asiatic society of Bengal-II: Fish drawings in Buchman-Hamiltion Zoological drawings, 22(3): 99–15.

Hora, S.L. (1926a). On a Peculia Fishing implements form Kangra Valley Punjab. *J. Asiat. Soc. Beng. N.S.* 21(1): 81–84 (1927).

Hora, S.L. (1935). Ancient Hindu conception of correlation between form and locomotion of fishes. *J.A.S.B. (Science)*, 1: 1–7.

Hora, S.L. (1948). Knowledge of the Ancient Hindus concerning fish and fisheries of India reference to fish in Arthasastra (Ca 300 B.C). *J.A.S. Bengal (letters) Kolkata*, 1(1): 7–10.

Hora, S.L. (1950). Knowledge of ancient Hindu concerning fish and fisheries of India II: Fishery, legislation on Asoka's Piller edict V 246 B.C., *Roy Asiat. Soc. Begal. Letters*, 26(1): 43–46.

Hora, S.L. (1951). Knowledge of ancient Hindu concerning fish and fisheries of India III: Matsyavinoda or a chapter on Angling in the Mansallosa by King Somesvara (1127 A.D.). *J.A.S. Bengal (Letters) Kolkata*, 17(2): 145–169.

Hora, S.L. (1953). Knowledge of ancient Hindu concerning fish and fisheries of India IV: Fish in the Sutra and Smriti Literature *J.A.S. Bengal (Letters) Kolkata*, 21(1).

Hora, S.L. (1956). Fish paintings in third Millennium B.C. from Nal. (Baluchistan) and their zoogepographical significance. *Mem. Ind. Mus., Kolkata*, 14(2): 78–84.

Hora, S.L. and Misra, K.S. (1938). Fish of Deolali Part 3: on two new Species and notes on some other forms. *J.B.N.H.S.*, 40(1): 35–50.

Hora, S.L. and Misra, K.S. (1941). Fishes of Satpura Range, Hoshangabad District. *R.I.M.*, 43(3): 361–374.

Hora, S.L. and Misra K.S. (1942). Fish of Poona Pt–II. *J.B.N.H.S.*, 43: 218–225.

Hora, S.L. and Saraswati, S.K. (1955). Fish in the Jatak tales. *J.A.S. Bengal (Letters) Kolkata*, 21(1): 13–30.

Hornell, J. (1942b). The Boats of the Ganges. *Mem. Asiat. Soc. Beng.*, 8: 173–198.

Hultzsh, E. (1925). Inscription of Asoka in Corpus Inscriptionum (Inscriptonum–indicarum), 1: 128.

Khan, Hamid (1946–47). A Fishery survey of the river Indus. *J.B.N.H.S.*, 46(3): 529–535.

Mackay, E. (1930). Painted pottery in Modern Sind: A survival of an ancient Industry. *J. Roy. Anth. Inst.*, 60.

Mackay, E. (1938). Further Excavations at Moherjordaro, Vol. I and II, Delhi.

Mahmood, S. and Rahimullah. M. (1947). Fish survey of Hyderabad State, IV: Fishes of the Nizamabad District. *J.B.N.H.S.*, 47(1): 102–110.

Majumdar, N.G (1934). Explorations in Sind. Mem. Arch. Surv. Ind. No. 48, Delhi.

Mashall, John (1931). Mohenjodaro and the Indus Valley culture, Vols I, II and III, London.

Menon, A.G.K. (1949). *Fishes of the Kumaon.*

Mosses, S.T. (1942). The fishery of Gujarat Coast. *J. Gujarat. Res. Soc.*, 4: 61–82.

Mosses, S.T. (1944). Report, Department of Fish, Baroda State (1942–43), Baroda.

Naidu, M.R. (1939). Report on a survey of the fisheries of Bengal, Government of Bengal, Department of Agricultural and Industry.

Nicholoson, F.A. (1915). Papers from 1819 relating chiefly to the development to the Madrah Fisheries Bureau, Bull. Madras Fisheries Bur, 1.

Radcliffe, W. (1921). Fishing from earliest times London.

Rahimullah, M. (1943c). Fish survey of Hyderabad State, II: Fishes of Hyderabad City and its suburbs. *J.B.N.H.S.*, 44(1): 88–91.

Rahimullah, M. (1944). A plan for the development of Fisheries in H.E.H. Nizam's dominions. Government Press. Hyderabad.

Rahimullah, M. (1944). Fish survey of Hyderabad State, III: Fishes of the Medak District. *J.B.N.H.S.*, 45(1): 73–77.

Rahimullah, M. (1946). Acclimatisation of exotic fish *Etroplus suratensis* (Bloch) in the Hyderabad State. *Proc. 33rd Ind. Sci. Cong.*, (Banglore, 1946). Pt. III. Abstract: 129.

Rao, H.S. (1956–57). History of our knowledge of the Indian fauna through the ages. *J.B.N.H.S.*, 54(2): 252–279.

Sewell, R.B.S. (1947). Memorandum of the proposed fishery research Institute. Govt. of India Press.

Sharma, R.S. (1986). The ancient India Delhi, p. 28.

Southwell, T. (1915). Report on fishery investigation in Bengal, Bihar and Orissa with recommendation for future work. Bull. Deptt. Fish, Bengal, Bihar, Orissa, p. 5.

Tampose, A. (1929). Report of the committee on fisheries in Madras. Govt. press Madras.

Thomas, H.S. (1870). Report on Pisciculture in South Canara, London, p. 1–67.

Thomas, H.S. (1873). *The Rod in India.*

Thomas, H.S. (1887). *Tank Angling in India*, p. 1–69.

Thomas, H.S. (1887). *Tank Fishing in India.*

Vats, M.S. (1940). *Excavation at Haryana*, Vol. I & II, Delhi.

Venkayya, V. (1906). Annual Report Archaeological Survey of India for 1903–04, p. 202–211.

Wolley, C.L. (1934). Ur. Excavations Vol. II (The Royal Cemetery).

"SPECIES INDEX"

A

Anabas scandens, 109

Arius sp., 12, 14

B

Bagaritus bagrius, 62, 116, 117

Bagarius bagarius, 65

Bagarius, 107

Barbus (Tor) *kludree*, 92, 109

Barbus curmuca, 62

Barbus nigrofaciatus, 94

Barbus sarana, 62

Barbus titteya, 94

Barillius spp., 116

Barillus benidelisis, 62

Botia, 10

C

Callichrous pabda (Hamilton), 45

Carassius carrasius, 91

Catla bunchanani, 109

Catla catla, 43

Cirrhina mrigala, 109

Clupea ilisha, 107

Clupisoma garua, 62

Crassius auratus, 94

Crassochilus, 10

Crocodilus porosus, 44

Ctenopharyngodon idella, 124

Cybium commersonii, 107

Cybium guittalum, 108

Cyprinion irregularis, 11

Cyprinion microphthalmum microphthalmum, 11

Cyprinion microphthalmum nikolskii, 11

Cyprinion milesi, 11

Cyprinion watsoni, 11

Cyprinion, 11

Cyprinus carpio, 123

Cyprinus communis, 123, 124

Cyprinus specularis, 123

E

Egypt, 15
Esomus, 62
Eurycantho latro, 13
Eutropiichthys vacha, 43, 62
Eutroplus suratensis, 108

G

Garra mullya (Sykes), 30
Garra, 10
Glyptothorax madrapatamum, 117
Glyptothorax, 10, 11
Gymmocorymba ternelzi, 94

H

Harpodon neherus, 108
Hemichromis bimaculatus, 94
Hyphessobrycon flammenus, 94
Hypophtalmichtrys molitrix, 124

L

Labeo boggut, 117
Labeo calbasu, 116, 117
Labeo dyocheilus, 116, 117
Labeo fembriata, 62, 64
Labeo fimbriatus (Cuvier & Valenciennes), 3, 117
Labeo gonius, 116
Labeo kontius, 117
Labeo rohita, 43, 45, 94, 116
Labeo spp., 2, 109
Labio dero, 116
Lates calcarifer, 108
Lates reticulates, 93
Lebister reticulates, 93

M

Mahseer, 10
Mastacembelus armatus (Lecepede), 30
Mastacembelus armatus, 62
Mollicenisia spherops, 94
Mugil corsula, 98, 108
Mugil oeur, 108
Mugil speigleri, 108
Mystus aor, 61, 116
Mystus cephalus, 118
Mystus seenghala, 116
Mystus spp., 117

N

Nemachilus sp., 10
Notopterus chitala, 62, 64

O

Ophicephalus gachuu, 94
Ophicephalus punctatus, 62
Ophicephalus sp., 106
Ophicephalus striatus (Bloch), 31, 62, 65
Oryzias melastigma, 94
Osphronemus goramy, 91

P

Pangasius pangasius, 64
Platinist gangetica, 45
Platypoecilus moculatus, 94
Polynemus paradisicus, 108
Polynemus tetradactylus, 108
Polynemus tetradoctylus, 64
Pongasius buchanani, 107
Pongassius, 116

Pterophyllum selare, 94

Puntius carraticus, 117

Puntius goninotus (tawes), 124-125

Puntius pangasius, 117

Puntius sophore (Hamilton), 44, 62

R

Rasbora doniconius, 94

Rasbora heteromorpha, 94

Rhinodon typics, 42, 43

Rita pavimentala, 117

Rita, 2, 12, 13

T

Tilapia mossambica, 94, 124

Tinca tinca, 91

Tanichthys albonubes, 94

Tor khudrae, 62, 117

Tor mahanodicus, 117

Tor mussulah (Saykes), 62, 117

Tor pituturo, 9, 10, 15, 116

Tor tor, 116, 117

V

Vallisneria, 46

Wallago attu, 2, 12, 31, 43, 45, 62, 65, 107, 116, 117

Xenextodon cancila, 62

Xiphophorus hellerii, 94

"SUBJECT INDEX"

A

Accidental transplantations, 94

Adharveda, 3

Aerial traps, 130

Air-breathing fish culture, 150

Air-breathing fish farming, 125

Alpara, 4

Amri, 2

Amri-Nal-Culture, 8

Ananda-machchha, 42, 47

Anaprasna, 4

Anathikamachhe, 50

Anattimika, 3

Andaja , 34

Angling, 95, 103

Animal systematics in Hindu, Buddhist and Jaina culture (600 B.C.-500 A.D.), 34

Anupa (frequenting marshy lands), 35

Arannaka (Wild), 3

Aranyakanda, 17, 31

Aristotle, 3

Aritr, 19

Aritra (Oar), 19

Art of angling, 60

Arthasastra, 3

Ashoka, 3

Ashoka's fishery legislation, 50

Ashoka's pillar Edict V, 49

B

Bag bets with fixed mouth, 131

Bag nets, 100

Baluchinstan, 1, 11

Beach-seine with peripheral pockets and without bag, 134

Beach-seine without peripheral pockets and bag, 133

Bhubaneshwar, 17

Boat seine, 134

Boats and ships in the Epics, 19

Boats in ancient India, 17

Boats in the British period, 110

Boats in Vedic period, 19

Brackishwater fish culture, 121

Brackishwater fisheries, 117

Brackishwater ponds and impoundments, 123

Brata, 4

Breeding of carps, 120

Breeding of Hilsa, 86

Breeding of other fishes, 83

Brhacchiras, 28

British colonial, 5

Bronze age period (5000 B.C.), 4

Buddha, 2

Buff-ware culture, 8

Buoyed traps, 129

C

Capt. R. Beavan (1877), 79

Cargo carriers, 112

Carp breeding, 83

Carp culture, 146

Carp farming, 149

Cast nets, 101

Casting the line, 66

Cats nets, 135

Ceta, 27

Chakra's classification of animals, 34

Chandragupta, 38

Chapadda (Hexapeds), 3

Chatupada (Quadrupeds), 3

Cilcima, 28

Classification of animals of Prasastapada, 35

Classification of animals of Susruta, 35

Coldwater fish culture, 122

Coldwater fisheries of hills streams, 119

Coldwater fisheries, 151

Composite fish culture, 125

Cooperative movements among fisherman, 143

Cover boats, 135

Culture fisheries, 120

Curing with salt, 106

Curing, 106

D

Description of the Pampa lake in the Ramayana, 32

Development of the inland fisheries resources, 122

Devpal, 3

Dinghis, 111

Dip nets, 102

Dipida (Bipeds), 3

Dr. Albert C.L.G. Gunther (1859-1870), 77

Dr. Cantor (1839-1860), 77

Dr. Francis Day (1826-1889), 78

Dr. J. Travis Jenkins (1908-1909), 80

Dr. John McClelland (1839-1841), 77

Dr. N. Annandale (1907), 80

Dr. P. Bleeker (1819-1878), 77

Dr. T.C. Jerdon (1835-1897), 79

Drag nets, 99

Drift nets, 101

Drive-in nets, 134

Dyumna (Raft), 19

E

Early vedic period (1500 B.C.-1000 B.C.), 21

Economics of retailers, 141

Economics of vendors, 141

Estuaries of north-west coast, 118

Evolution of fishery laws in India, 53

Exotic fish culture, 121

Falling nets, 135

F

Fast days in relation to fishery conservation, 53

Fish and fisheries in Sutra and Smiriti literature (600 B.C.-200 A.D.), 22

Fish and fisheries in the Gupta period (350 A.D.-550 A.D.), 56

Fish and fisheries in the Ramayana and ancient India (200 B.C.-700 A.D.), 30

Fish and fisheries on Ashoka pillar edict V (246 B.C.), 48

Fish and fishes in the Jataka tales (380 B.C.-500 A.D.), 42

Fish as food, 23, 36, 47

Fish barrier, 129

Fish bite, 66

Fish farming in integrated system, 126

Fish fauna of India studied during the period (1938-1950), 82

Fish forbidden as food, 24

Fish hooks from the Indus valley, 13

Fish markets, 138

Fish omens, 47

Fish pastes, 106

Fish pot, 129

Fish preservation and processing in the British India, 105

Fish protection, 25

Fish remains of Harappa culture, 12

Fish scales, 4

Fish tackle, 63

Fish traps, 129

Fisheries administration and management, 87

Fisheries in Bengal, 57

Fisheries in Harappan Chalcolithic culture, 12

Fisheries in the British colonial period, 66

Fisheries in the medieval period, 57

Fisheries in the Sangam literature, 56

Fisheries of brackishwter lagoons, 119

Fisheries of freshwater reservoirs and lakes (Lacustrine fisheries), 118

Fisheries of freshwater rivers and canals, 116

Fisheries science in the Buddhist period and the Magdghon empire (600 B.C.-232 B.C.), 38

Fisherman, 141

Fishing by hand, 96

Fishing in the British period, 94

Fishing villages, 48

Four decades of Central Inland Fisheries Research Institute, 143

Freshwater po[illegible] ꞁanks and swamp res[illegible]s, 123

Freshwater riverine forms, 61

Full moon periods in relation to fishery conservation, 52

G

Ganga puputake, 50

Gesner, 3

Ghara (Domestic), 3

Gill net, 102, 136

Gross Domestic Product (GDP), 154

Ground bait, 65

H

Habits and instincts of fishes in Jataka tales, 46

Harappa, 2, 12, 13, 16

Hexapeds, 3

Hilsa fisheries in Madras and Bombay, 86

Hilsa fisheries, 151

Hook, 64

Hunting, 95

I

Identification of fishes, 26

Indian fishery, a future vision, 154

Indian salmon group, 108

Indus valley, 1

Indus valley civilization, 13

Inland fish and fisheries in the Jataka sculptures (200 B.C.), 54

Inland fish flows, 139

Inland fishing gear, 127

Interspecific hybrids, 121

J

Jala, 3

Jalacocera (aquatic), 3

Jangala (dwelling in dry areas), 35

Jarayuza, 34

K

K.G. Gupta (1906-1908), 80

Kalidasa, 2

Kanamahamaccha, 43

Kaphatasykae, 50

Kaulli culture, 8

Kautilya Arthasastra, 38

Kautilya, 3

Kerala backwaters, 119

Kravyabhuja (Carnivorous), 35

Ksura-nasika, 44

Kumbhila, 44

Kyoto declaration, 155

L

Lacocera (terrestrial), 3

Lantern nets, 135
Large fishing boats, 111
Later Vedic period (1000 B.C.-600 B.C.), 21
Line fishing, 128
Line, 63
Lion-man, 2
Live fishes group, 108
Lord jagannatha, 18

M

Mahabharat, 4
Mahakapa jataka, 55
Mahamukhamaccha, 43
Mahanadi system, 118
Mahasakari, 29
Mahasalka, 27
Makara, 44
Makura medallion, 55
Manpower development, 152
Manuscript drawings of fish in the library of A.S.B. (Mackenzie collection, 1784-1814), 72
Mardura, 28
Marine and estuarine forms, 60
Market intermediaries, 139
Marketing of fish in British period, 107
Marketing of fish, 137
Medieval period, 5
Mesolithic period, 7
Mesopotamia, 11, 15
Methods of catching fish, 47
Methods of fishing at Nal, 12
Migration of Hilsa, 86
Mirror carp, 124
Miscellaneous, 95, 104
Mohanjodaro, 2, 4, 17
Mr. Blythe (1841-1862), 78
Mr. Grote and Col. Strochey (1876), 78
Mr. T. Southwell (1911-1918), 81
Munja-rohita, 45

N

Naga jataka, 54
Nal, 2, 8
Nalamina, 30
Nanda dynasty, 38
National aquacultural research, 159
Navaja, 19
Neolithic period (3500 B.C.-1700 B.C.), 8
Net sinkers, 16
Nets without closing string with peripheral pockets, 135
Netting, 95, 99
Nundara, 2

P

Paddy-cum-fish farming, 126
Pal dynasty, 3
Paleolithic, 7
Palini, 3
Parastapada, 3
Pathin,a 45
Pavusa, 45
Persia, 11

Playing a fish, 66
Plunge basket, 135
Post-Vedic literature, 3
Post-Vedic period, 22
Pot-shreds from Harappa, 13
Pre-Harrapan peasant communities, 7
Preservation in mustard oil, 106
Price spread in terminal sale through retailers, 140
Price spread in terminal sale through vendors, 141
Prohibition of tank fishing in relation to fishery conservation, 53
Prthuloma, 45
Punjab, 11
Purse nets, 136
Push net, 131

Q

Quetta culture, 8
Quill Float, 65

R

Raft trap, 130
Rajiva, 29
Ramayan, 3
Redware culture, 8
References of boats in the Sanskrit and Pali literature, 18
Religious injunctions pertaining to fish and fishing, 25
Remains of carp, 13
Research programmes, 144
Reservoir fisheries, 152
Retail markets, 140
Rig veda, 3
Rod, 63
Rohita, 29, 31, 43
Rowing and roring skiffs, 111
Ruper, 2

S

Sahasradomstra, 28
Sakula, 30
Sales of fish, 48
Samhitas, 3
Samkujyamacche, 50
Sanakra, 3
Sangam age, 3
Saphori, 26
Sarpasira, 27
Sasalka, 29
Satabali, 26
Satapada, 3
Scale carp, 124
Science net, 99
Scientific analysis of Ashoka's Laws, 52
Sewage fed fish farming, 125
Silundi gangetica, 107
Sind, 2, 11
Slow net on anchor, 132
Slow net on stakes, 132
Smhatundra or Simhatundaka, 29
Socio economic development of fisherman communities in India, 140
Somesvara's advice on fishing, 65

Special nets used to catch Hilsa, 100

Striking a fish, 666

Stupa at Sanchi, 18

Sun drying, 105

Surrounding gear, 134

Susruta contribution to Ichthyology (400 B.C.), 36

Susruta, 3

Sustras, 23

Susu, 45

Suvannavonna maccha, 44

Svedoja, 35

Syria, 11

System of freshwater fish farming, 125

T

Timingila jataka, 55

Torkle, 65

Transplantation of fishes during British period, 89

Transplantation of food fishes, 89

Transplantation of game fishes, 91

Transplantation of larvicidal fishes, 93

Transplantation of ornamental fishes, 93

Transportation of fishes in independent India, 123

Trapping, 96

Traveling houseboats and ceremonial barges, 112

Trawl nets, 99

U

Uda Jataka, 55

Udvija, 35

Umasvati classification, 36

Umravati, 3

Upanishad, 3

Usmaja, 35

V

Valaja, 45

Varmi, 28

Varondah net, 131

Vasistha Dharmasutra, 23

Vedaveyake, 50

Velomoccha, 44

Vindusarowara, 18

Vishnu, 2

Vishnugupta, 38

W

W.A. Sykes (1824-1831), 76

Wholesale intermediaries, 139

Wholesale markets, 140

Y

Yajurveda, 3

Yukti Kapatau, 18

Special nets used to catch Hilsa, 104

Sri kanta fish, 66

Stupa at Sanchi, 19

Stupefying, 108

Surrounding gear, 134

Susruta contribution to ichthyology (600 B.C.), 36

Susruta, 3

Sustas, 22

Sutu, 45

Suvarnavarna macha, 44

Svedaja, 35

Syria, 11

System of freshwater fish farming, 25

T

Timingila jataka, 55

Tackle, 63

Transplantation of fishes during British period, 85

Transplantation of food fishes, 89

Transplantation of game fishes, [illegible]

Transplantation of larvicidal fishes, 93

Transplantation of ornamental fishes, 93

Transportation of fishes in independent India, 123

Trapping, 96

Trawling in reservoirs and continental ranges, 117

Trawl nets, 90

U

Uddal jataka, 55

Udyog, 38

Unimproved classification, 26

[illegible]

Upanishad, 3

Usmaja, 36

V

Vatsja, 45

Varmi, 23

Varamin net, 131

Vasistha Dharmasutra, 22

Vetarvyaloe, 51

Vehinnacetta, 44

Vindusangraha, 18

Vishnu, [illegible]

Vishnugupta, 33

W

W.A. Sykes (1821-1831), [illegible]

Wholesale intermediaries, 139

Wholesale markets, 140

Y

Yajurveda, 3

Yuktikalpataru, 18